THE OCEAN BASINS: THEIR STRUCTURE AND EVOLUTION

THE OCEANOGRAPHY COURSE TEAM

Authors
Evelyn Brown (*Waves, Tides, etc.; Ocean Chemistry*)
Angela Colling (*Ocean Circulation; Seawater (2nd edn); Case Studies*)
Dave Park (*Waves, Tides, etc.*)
John Phillips (*Case Studies*)
Dave Rothery (*Ocean Basins*)
John Wright (*Ocean Basins; Seawater; Ocean Chemistry; Case Studies*)

Designer
Jane Sheppard

Graphic Artist
Sue Dobson

Cartographer
Ray Munns

Editor
Gerry Bearman

Other titles in the oceanography series

The Ocean Basins: Their structure and evolution
Ocean Circulation
Waves, Tides and Shallow-water Processes
Marine Biogeochemical Cycles
Case Studies in Oceanography and Marine Affairs
Biological Oceanography: An introduction

Cover illustration: Satellite image of a coccolithophore (*Emiliania huxleyi*) bloom in the English Channel off the south coast of Cornwall, 24 July 1999. (*Andrew Wilson and Steve Groom.*)

THE OCEAN BASINS: THEIR STRUCTURE AND EVOLUTION

PREPARED BY JOHN WRIGHT AND DAVID A. ROTHERY FOR THE COURSE TEAM
SECOND EDITION REVISED FOR THE COURSE TEAM BY DAVID A. ROTHERY

Butterworth-Heinemann
An imprint of Elsevier
books.elsevier.com

Published by

Butterworth–Heinemann (an imprint of Elsevier)
Linacre House, Jordan Hill, Oxford OX2 8DP, UK
30 Corporate Drive, Suite 400, Burlington, MA 01803, USA

in association with

The Open University
Walton Hall, Milton Keynes
MK7 6AA
United Kingdom

First published 1989. Second edition 1998. Reprinted 2001, 2002, 2007.

Edited, designed and typeset by The Open University.

Printed and bound in Singapore under the supervision of MRM Graphic Ltd. Winslow, Bucks.

This book forms part of an Open University course S330. Details of this and other Open University courses can be obtained from the Student Registration and Enquiry Service, The Open University, PO Box 197, Milton Keynes MK7 6BJ, United Kingdom: tel. +44 (0)845 300 60 90, email general-enquiries@open.ac.uk

http://www.open.ac.uk

British Library Cataloguing in Publication Data

A catalogue record for this book is available from the British Library.

Library of Congress Cataloging in Publication Data

A catalogue record for this book is available from the Library of Congress.

ISBN 978-0-7506-3983-5

10 9 8 7 6 5 4 3 2 1

2.5

CONTENTS

ABOUT THIS VOLUME

This is one of a Series of Volumes on Oceanography. It is designed so that it can be read on its own, like any other textbook, or studied as part of S330 *Oceanography*, a third level course for Open University students. The science of oceanography as a whole is multidisciplinary. However, different aspects fall naturally within the scope of one or other of the major 'traditional' disciplines. Thus, you will get the most out of this Volume if you have some previous experience of studying geology, geochemistry or geophysics. Other Volumes in this Series lie more within the fields of physics, chemistry and biology (and their associated sub-branches).

Chapters 1 to 4 describe the processes that shape the ocean basins, determine the structure and composition of oceanic crust and control the major features of continental margins. Today's ocean basins are geologically ephemeral features, and these Chapters show why. Chapter 5 deals with the 'hot springs' of the deep oceans that result from the circulation of heated seawater through oceanic crust. This phenomenon was not even suspected until the mid-1960s and was not confirmed by observation until some years later. Since then, many people have seen the striking photographs of 'black smokers' at ocean ridges. Chapter 6 summarizes the main patterns of sediment distribution in the ocean basins and shows how sediments can preserve a record of past climatic and sea-level changes. Finally, Chapter 7 considers the role of the oceans as an integral part of global chemical cycles.

You will find questions designed to help you to develop arguments and/or test your own understanding as you read, with answers provided at the back of the Volume. Important technical terms are printed in **bold** type where they are first introduced or defined.

ABOUT THIS SERIES

The Volumes in this Series are all presented in the same style and format, and together provide a comprehensive introduction to marine science. *Ocean Basins* deals with the structure and formation of oceanic crust, hydrothermal circulation, and factors affecting sea-level. *Seawater* considers the seawater solution and leads naturally into *Ocean Circulation*, which is the 'core' of the Series. It provides a largely non-mathematical treatment of ocean–atmosphere interaction and the dynamics of wind-driven surface current systems, and of density-driven circulation in the deep oceans. *Waves, Tides and Shallow-Water Processes* introduces the physical processes which control water movement and sediment transport in the nearshore environment (beaches, estuaries, deltas, shelves). *Ocean Chemistry and Deep-Sea Sediments* is concerned with biogeochemical cycling of elements within the seawater solution and with water–sediment interaction on the ocean floor. *Case Studies in Oceanography and Marine Affairs* examines the effect of human intervention in the marine environment and introduces the essentials of Law of the Sea. The two case studies respectively review marine affairs in the Arctic from an historical standpoint, and outline the causes and effects of the tropical climatic phenomenon known as El Niño.

Biological Oceanography: An Introduction (by C. M. Lalli and T. R. Parsons) is a companion Volume to the Series, and is also in the same style and format. It describes and explains interactions between marine plants and animals in relation to the physical/chemical properties and dynamic behaviour of the seawater in which they live.

CHAPTER 1 INTRODUCTION

Humans have made use of the seas throughout history and the unrecorded past for activities such as fishing, transport, trade and warfare. The first sea-going craft may have been developed from the river boats of the oldest known civilizations (Figure 1.1). Up to the fifteenth century there was little attempt at systematic exploration of the oceans, though some epic voyages were made, especially in the Atlantic and Indian Oceans from Europe, North Africa and the Middle East, and over substantial areas of the western Pacific by Polynesians and Melanesians. Many regions were first explored by people other than Europeans. For example, in the early 1400s Chinese explorers reached the east coast of Africa by sea before the Portuguese, though their exploration was not followed by any attempt at colonization or trade.

Figure 1.1 A model of an ancient Egyptian river craft of about 2000 BC, from a tomb at Thebes.

1.1 MAPPING THE OCEANS

In the fifteenth century, Europeans began to dominate exploration of the oceans. Patrons such as Prince Henry the Navigator of Portugal encouraged exploration as much for its own sake as to find new trade outlets. During the fifteenth to eighteenth centuries, Columbus, Magellan, Cook and others advanced seafaring techniques and our knowledge of global geography – and brought in their wake world-wide European colonization.

Early voyages of exploration were perilous, not least because of the difficulties of navigation and position-fixing out of sight of land. The few rudimentary instruments available to the navigator were limited to finding only the latitude of the ship at sea. No simple reliable means of finding the longitude was available until the Englishman John Harrison developed an accurate and reliable chronometer in the eighteenth century.

In view of these difficulties, Halley's map (Figure 1.2) is particularly remarkable. This shows that by 1701 the shape of the Atlantic Ocean basin had been determined, which was certainly not the case barely 200 years before (Figure 1.3). Halley's map shows coastlines quite accurately with respect to longitude (which he determined by painstaking observations of the passage of the Moon in front of stars, and eclipses of Jupiter's satellites), and also contours (isogons) of the amount of angular divergence between magnetic and true (or geographic) north. Halley proposed that this angular divergence, which he called 'variation', should be measured by mariners as a contribution to determining their position.

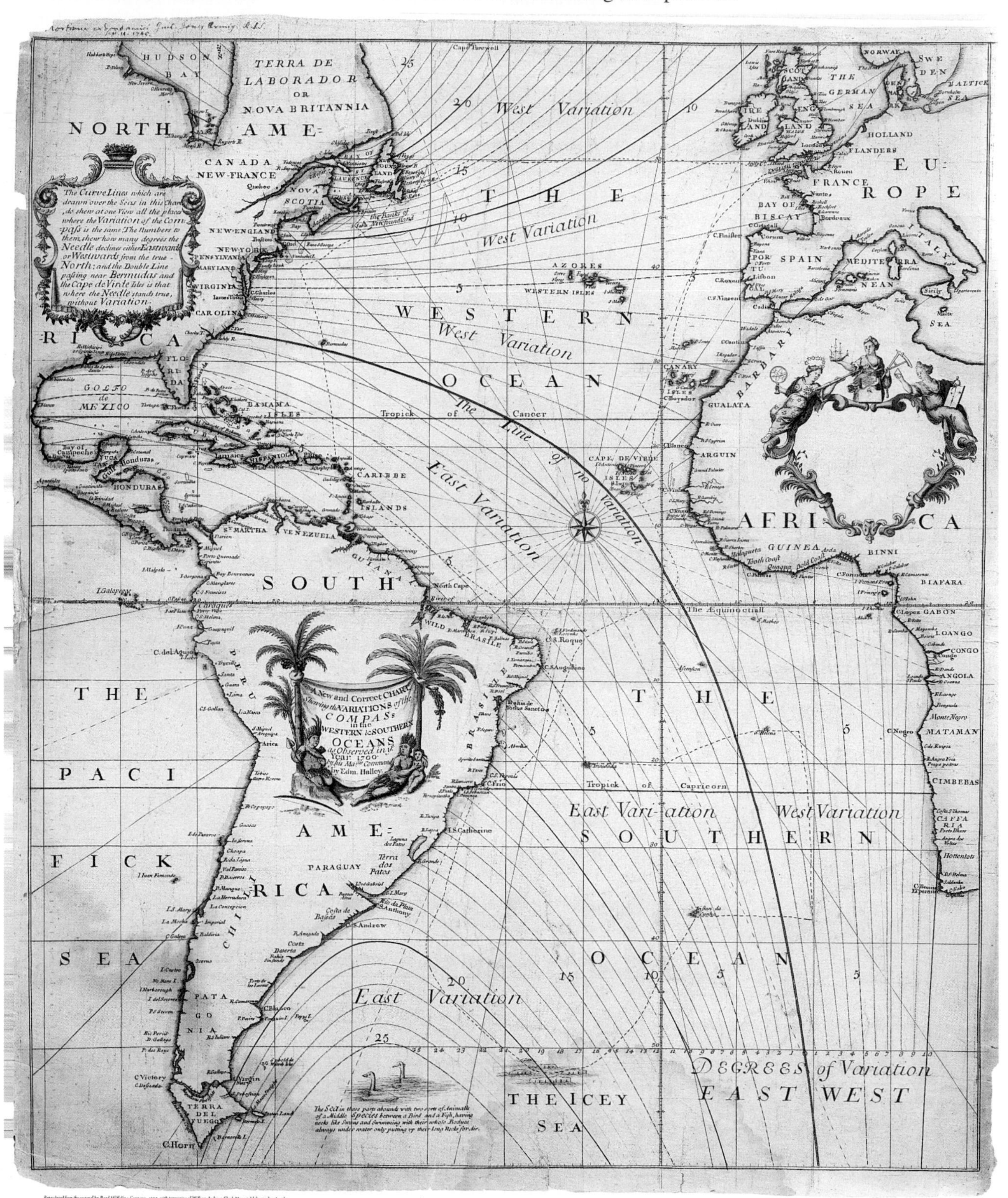

Figure 1.2 Map by Edmund Halley, 1701, produced using data collected during two voyages on the *Paramore* (sometimes spelt *Paramour*), 1698–1700. The curved lines are isogons showing the angle between magnetic and true (or geographic) north. *Source*: N. J. W. Thrower (1972) *Maps and Man*, Prentice-Hall.

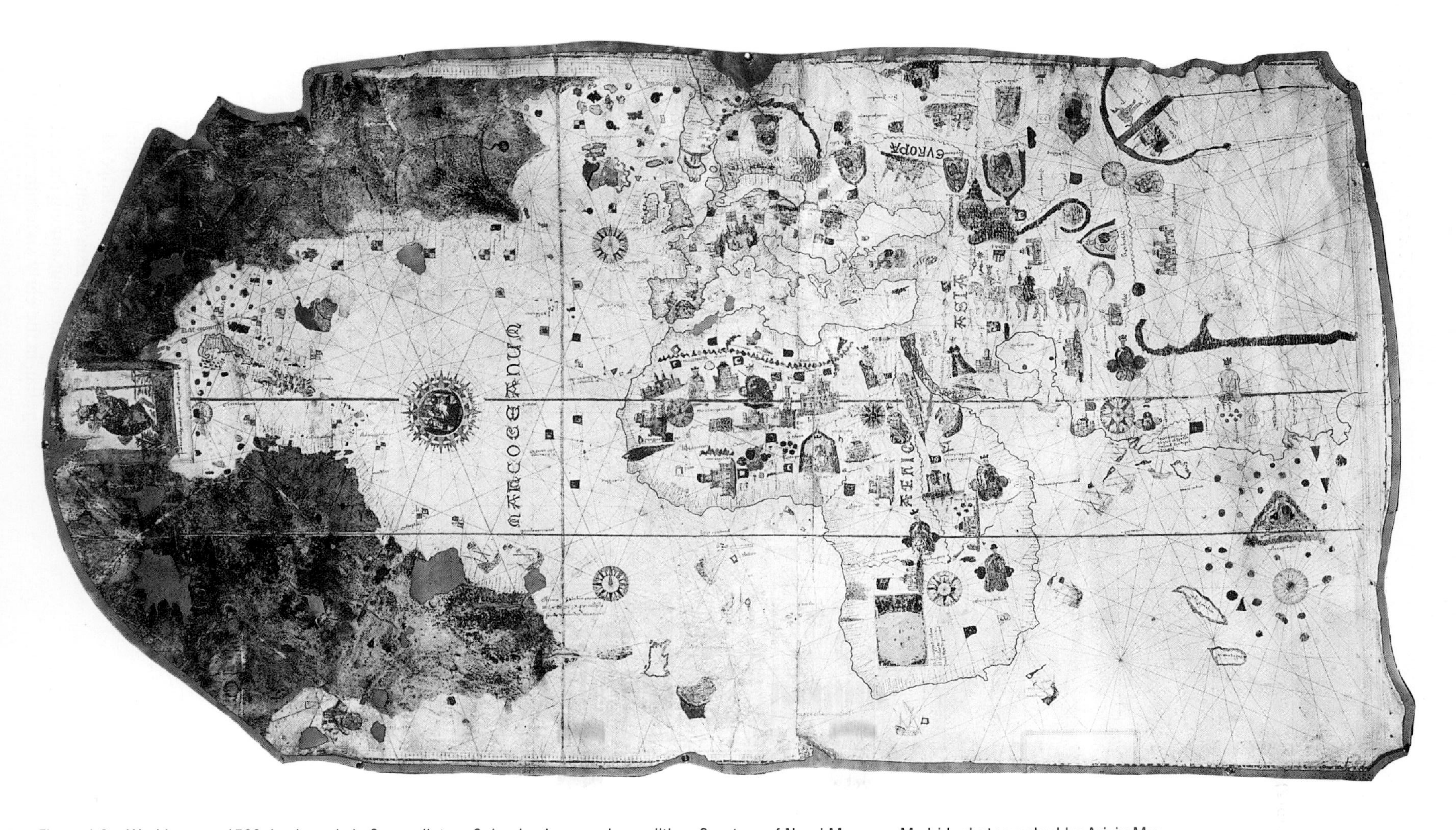

Figure 1.3 World map, *c.*1500, by Juan de la Cosa, pilot on Columbus's second expedition. Courtesy of Naval Museum, Madrid; photographed by Arixiu Mas.

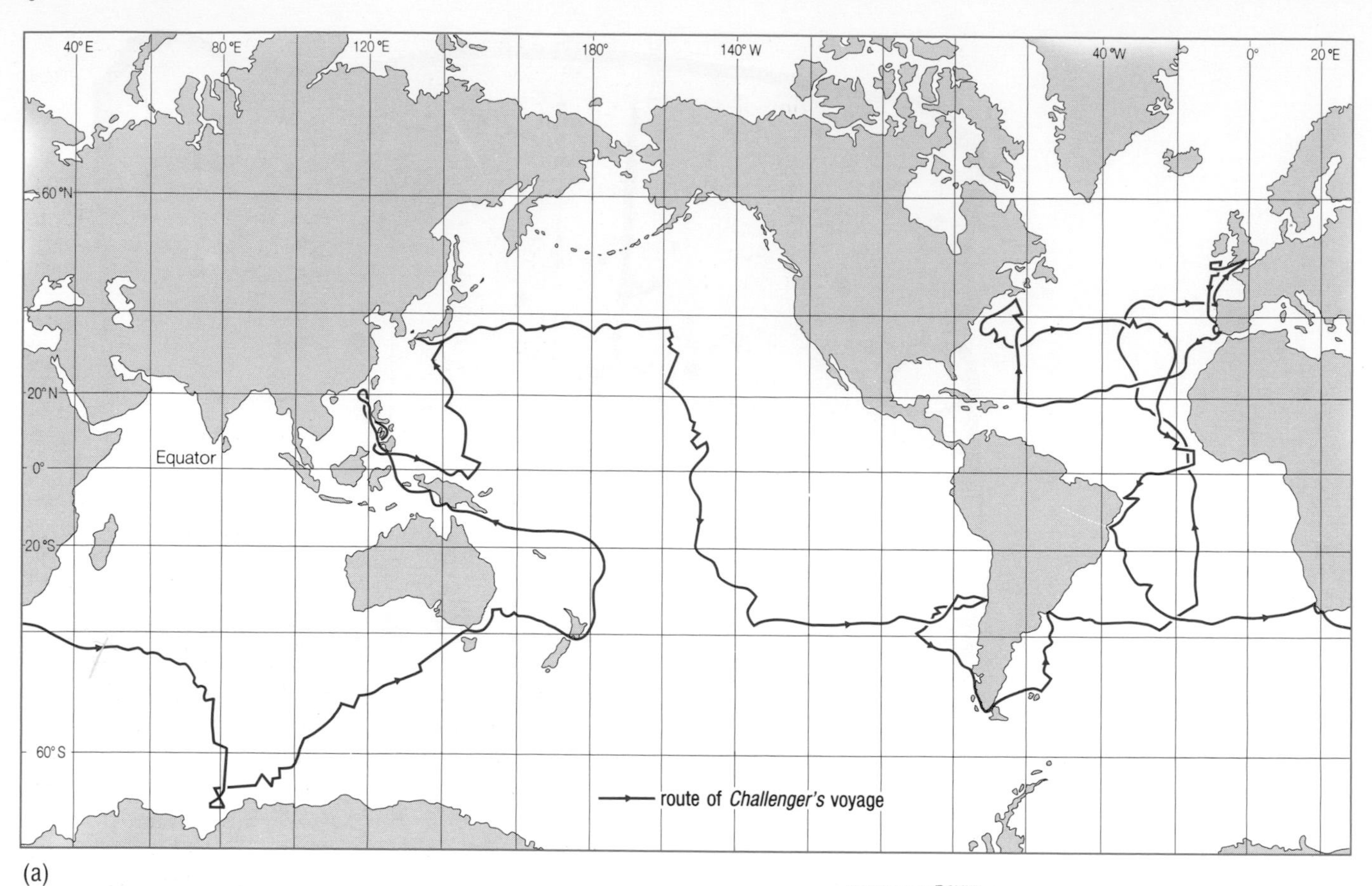

(a)

Figure 1.4 (a) The route of HMS *Challenger*, 1872–76.
(b) HMS *Challenger*, 1872. She was a steam-assisted corvette of 2306 tons.

(b)

The systematic gathering of oceanographic information was at first almost entirely concerned with geographical exploration and the preparation of coastal charts, so that knowledge of the oceans advanced little beyond the lore of seamen and fishermen. Very few investigations of the deep oceans were undertaken, but among the earliest was an unsuccessful attempt by Magellan in the early sixteenth century to sound the floor (see Figure 1.7) of the central Pacific, which proved too deep for his sounding lines. These studies increased during the nineteenth century, culminating in the *Challenger* expedition of 1872–76, mounted by the British Government to collect oceanographic information (Figure 1.4).

The huge quantity of data of all types collected during *Challenger's* voyage was published in 50 volumes and may be considered to mark the birth of scientific study of the oceans. Other aspects of oceanographic research were also developed during this period. Until the nineteenth century, mariners were aware of ocean currents principally because their voyages were delayed or speeded up by them. For example, monsoonal reversals of the Somali Current, off East Africa, were documented by Arabs as long ago as the ninth century. Changes in the colour of the sea and the appearance of seaweeds far offshore (as in the Gulf Stream) provided further evidence of ocean currents. However, the difficulties of navigation in the open oceans meant that the velocity of ocean currents could not be related easily to fixed reference points in the way that the better known coastal currents could be related to land features.

1.1.1 NAVIGATION

The problems of navigational accuracy experienced by the early explorers have been largely overcome by modern technology, but it remains easier to navigate within sight of land than in the open oceans. Coastal navigation depends on features being visible (or otherwise detectable) from the ship. Positions are fixed by taking two or more bearings on identifiable points. A bearing may be taken by compass, or by noting when two navigational marks come into alignment as seen from the vessel. By day, coastal navigational marks include topographic features (such as headlands), prominent buildings, radio masts, and deliberately placed beacon towers or posts. Channels and hazards (such as shoals and wrecks) may be marked by buoys. By night, many deliberately placed navigational marks are indicated by lights that flash in distinctive sequences, and whose colour may change from white to red or green depending on the observer's bearing. Factors limiting visibility (apart from bad weather and rough seas) are the Earth's curvature, the height of the observer, and the height of the object (Figure 1.5).

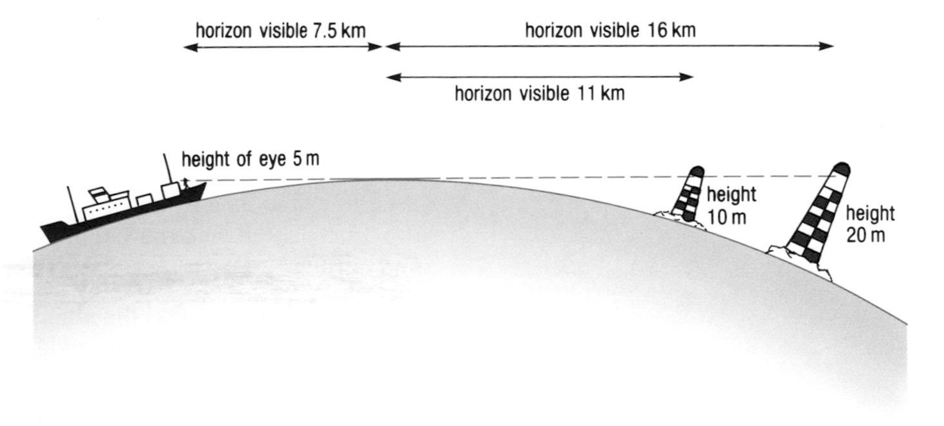

Figure 1.5 Visibility and the curvature of the Earth (not drawn to scale). The maximum distance at which an object can be seen directly (the 'dipping range') is the sum of the distances from observer to horizon and from horizon to object. The distance to the horizon depends on height. An observer 5 m above sea-level has a horizon at 7.5 km, and can see lights 10 m and 20 m above sea-level at distances of 7.5 + 11 = 18.5 km and 7.5 + 16 = 23.5 km, respectively.

In bad weather or at night, radar (*ra*dio *d*etection *a*nd *r*anging) is useful. Radio waves (at about 3000–10 000 MHz*) are transmitted from the ship in a rotating beam and reflected from objects in its line of sight. The time

*MHz = megahertz. The hertz (Hz) is the standard unit of frequency, which is 1 cycle per second. 1 MHz = 10^6 Hz and 1 kHz = 10^3 Hz.

between the transmission of the signal and reception of the reflection gives the distance, and the bearing is given by the direction from which the reflection arrives. The time between transmission and reception is very short, because radio waves are a form of electromagnetic radiation and travel at the speed of light ($\sim 3 \times 10^8\,\mathrm{m\,s^{-1}}$).

The accuracy of visual-fixing methods is of the order of tens to hundreds of metres, but these methods are limited to the line of sight and so have short ranges. The effective range of radar is similar, though it can sometimes 'see' beyond the horizon. Longer-range radio navigation systems use networks of land-based transmitters emitting pulsed or continuous radio signals with frequencies from 10 kHz to 2 MHz, which can travel some way round the curvature of the Earth. The ship's receiver measures the interference patterns set up between the radio waves from different stations, and uses this information to calculate the ship's position. The range and accuracy of such systems vary. For example, *Loran-C* covers the north-east Atlantic and the North Sea from a chain of stations stretching from Scandinavia to France and Ireland. It has a maximum range of about 2000 km, and a positional accuracy of ± 100 m in optimum conditions (when the range is less than 300 km) (see Table 1.1).

The classical method of navigation used in the open oceans is celestial navigation, otherwise known as astro-navigation. The angles of the Sun, Moon, planets and stars above the horizon are measured (using a sextant) at a precisely recorded time. The positions of these bodies in the sky can be looked up in tables (manually or using a computer), and an observation of any one of them gives a position line on which the ship is situated. When two or more position lines have been found, they should cross at a point defining a unique location, or 'fix', for the geographical position of the ship. Even under ideal conditions, celestial navigation is accurate only to about half a kilometre. Use of a sextant on a moving ship demands skill, and relatively laborious and time-consuming calculations are involved if there is no computer available. Observations are possible only when both the horizon and the reference body are visible. Both the Sun and/or Moon and the horizon may be visible during daytime, but at night the horizon is not easily visible, so measurements using the planets or stars usually have to be made at twilight, when the horizon can be seen.

Celestial navigation can obviously not be done continuously, so a ship's position at any time after the last celestial (or coastal) fix has to be estimated by 'dead reckoning'. The vessel's heading and speed through the water are used to extrapolate from the last fix to the current position, taking account of the time elapsed, and the estimated rate of drift off-course caused by winds and currents.

Celestial navigation was the mainstay of mariners for centuries, but now navigational satellites can provide positions to within ± 100 m virtually anywhere in the open oceans, regardless of weather conditions. Satellite navigation was developed by the US Navy and first came into civil use in 1967. The modern set-up is called the Global Positioning System, or GPS, and uses an impressive array of satellites (Figure 1.6). Each GPS satellite continually transmits accurately timed radio signals including precise information on its own position in space, which can be picked up by a receiver small enough to be hand-portable. Knowledge of the times of transmission and receipt of these signals, and of the speed of light, allows the receiver's software to determine the range to each GPS satellite. Ranges to

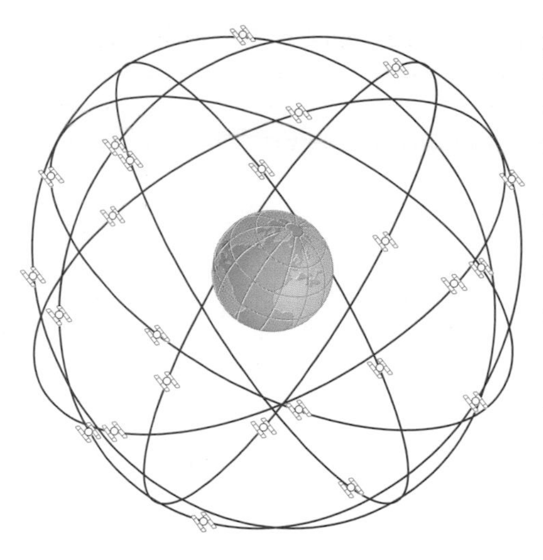

Figure 1.6 The GPS (Global Positioning System) satellite array, consisting of 24 satellites arranged in six orbital planes, each inclined at 55° to the Equator. The satellites are in circular orbits at a height of 20 200 km, with 12-hour orbital periods. At least four satellites are always 'in view' from a receiver anywhere on Earth.

just three of these are sufficient to fix the position of the receiver, though six or more are used when possible for greater precision. In this way, the position of a ship can be fixed to within ± 100 m. This remarkable accuracy is actually much poorer than the system is capable of, because the signals are deliberately degraded for civil use. US military receivers have decryption codes that can extract fixes accurate to the nearest metre. For detailed non-military survey work, it is worth using a ship-board receiver in conjunction with a fixed receiver on land; this so-called 'differential GPS' allows the ranging errors to the satellites to be calculated and corrected for, giving a ship's position to ± 10 m.

Precise navigation is vital to many sorts of oceanographic study. As a matter of principle, more than one navigational method is commonly available in case of instrument failure. Although GPS is becoming the standard technique for research, commercial and military navigation, it seems likely that the older methods will never be abandoned entirely. Table 1.1 summarizes the range and accuracy of the various navigational methods described in this Section.

Table 1.1 The range and accuracy of some navigational methods at sea.

Type	Range	Accuracy
coastal (visual and radar)	line of sight (up to about 50 km)	± 10–200 m
radio navigation systems	variable, e.g. *Loran-C* (NE Atlantic, up to about 2000 km offshore)	± 100 m–1 km
celestial navigation	world-wide	± 0.5–10 km
satellite navigation (GPS)	world-wide	± 100 m or less
differential GPS	word-wide	± 10 m

Figure 1.7 The Leadsman. Line soundings were the only means of depth measurement until the early twentieth century. (*Artist*: Arthur Briscoe)

1.1.2 DEPTH MEASUREMENT

Contoured topographic maps of land areas are based on the measurement of height *above* a sea-level datum. In the oceans, contoured bathymetric maps are based on measurements of depth *below* a sea-level datum. Accurate positioning is of course essential for any height or depth measurement.

Before the development of echo-sounders, knowledge of ocean depths was limited to relatively few isolated soundings, which were often of doubtful accuracy. These soundings were made by lowering a weighted rope or steel plumb-line until the weight reached the sea-floor. The depth was assumed to be the length of cable let out. In shallow water, soundings could be made relatively rapidly (Figure 1.7), but in the deep oceans it would take several hours to lower and raise the sounding line. During this time, currents and winds could move the ship from the position where the weight was lowered, so the line ended up being far from vertical and the depth would be grossly overestimated. Furthermore, in deep water it was difficult to determine when the weight reached bottom, as the long length of cable was much heavier than the weight itself. An indication of the difficulties can be gauged from the fact that in their three-and-a-half-year expedition the *Challenger* scientists achieved only some 300 deep-water soundings.

Echo-sounders were first used on oceanographic expeditions in the 1920s. Hundreds of depth measurements could be made in a matter of days, and it became possible to produce accurate bathymetric maps of major features such as the Mid-Atlantic and the Carlsberg Ridges, with their axial rift valleys.

Echo-sounding works on principles analogous to radar, except that acoustic or sound pulses (generally about 5–200 kHz) are used instead of radio waves (which cannot penetrate water very far). This is an example of **sonar** (*so*und *n*avigation *a*nd *r*anging), and can provide rapid, continuous and accurate depth measurement. The ship transmits a series of sound pulses through the water, which are reflected from the sea-floor and received as echoes. The interval between transmission and reception of the echo provides a measure of depth, using the known speed of sound in seawater. If pulses are transmitted at short intervals, reflections from the sea-floor can be recorded graphically to give a continuous profile (Figure 1.8). Under ideal conditions, an accuracy of within a few metres is possible even in deep water. However, there are several sources of inaccuracy. The most obvious is that surface waves may move the ship up and down through several metres. More importantly, the speed of sound in seawater can vary by about ± 4% according to variations in temperature, salinity and pressure. There are tables for different areas of the world's oceans, which can be used to make first order corrections of echo-sounder records (normally standardized to 1500 m s^{-1}). Inaccuracies in these tables can lead to appreciable errors in depth measurements. For example, a 1% error in the sound velocity means a 30 m depth error in 2 km of water.

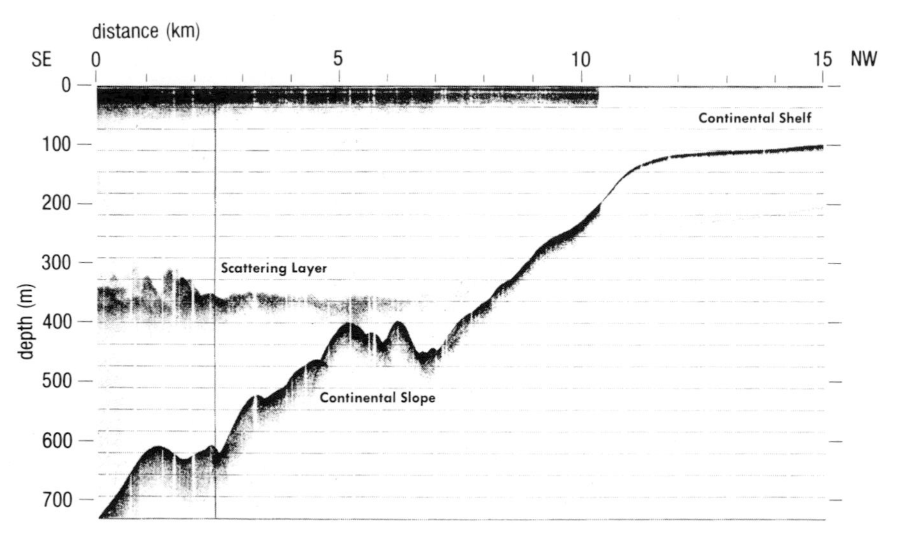

Figure 1.8 An echo-sounder record, giving a bathymetric profile of the edge of the eastern USA continental shelf. The slopes appear steep because the vertical exaggeration is × 12. The scattering layer probably represents a concentration of organisms and biological debris, and shows that echo-sounding has other applications.

Inaccuracies also result from the low resolving power of some echo-sounders. This is due to the divergence, at angles of about 30°, of the sound pulses as they are transmitted downwards, so that in deep water they cover large areas of ocean floor – *c.* 1.5 km across at 3 km depth (Figure 1.9(a)). Bathymetric variations within such an area will not be resolved, and the depth that is recorded could be shallower than the depth directly beneath the ship (Figure 1.9(b)).

Echo-sounders with higher resolution, using beam angles of a few degrees, have to be much bulkier in order to give a narrower beam, so they are not used except by survey vessels. Furthermore, a large angle of divergence is necessary for routine navigational purposes, to allow for use on a rolling ship in heavy seas. Narrow-beam instruments would be of little use in such situations, but are of course invaluable for survey work.

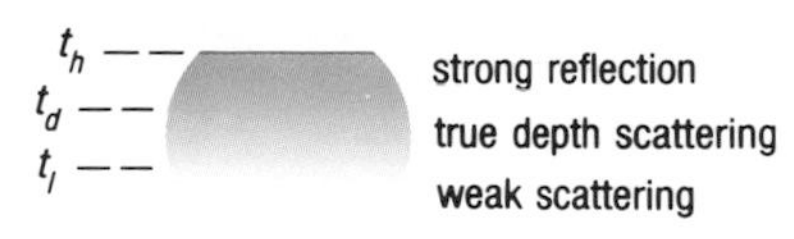

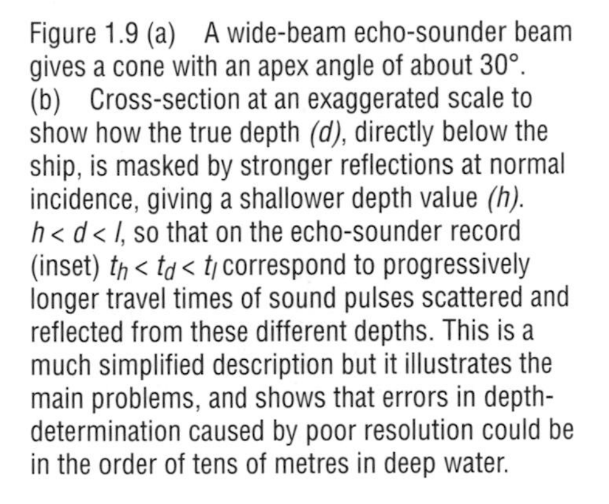

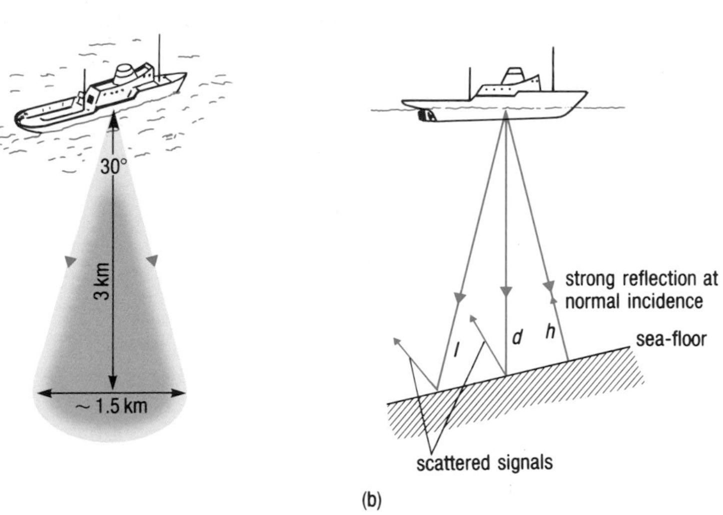

Figure 1.9 (a) A wide-beam echo-sounder beam gives a cone with an apex angle of about 30°. (b) Cross-section at an exaggerated scale to show how the true depth *(d)*, directly below the ship, is masked by stronger reflections at normal incidence, giving a shallower depth value *(h)*. $h < d < l$, so that on the echo-sounder record (inset) $t_h < t_d < t_l$ correspond to progressively longer travel times of sound pulses scattered and reflected from these different depths. This is a much simplified description but it illustrates the main problems, and shows that errors in depth-determination caused by poor resolution could be in the order of tens of metres in deep water.

QUESTION 1.1 Examine Figure 1.8 and estimate the reliability of the data on position and depth shown on the profile for (a) the edge of the continental shelf, and (b) the lowermost 'bump' on the continental slope.

Question 1.1 illustrates the important point that in surveying the oceans it may be necessary to tolerate errors of greater magnitude than are normally acceptable on land, where mislocating certain topographic features with respect to the line of, for example, a proposed road by even a few metres (let alone a couple of kilometres), would be unthinkable. However, positional accuracy needs to be very high indeed when siting a well-head on the sea-floor, or in detailed scientific surveys.

1.2 MAPPING THE OCEAN FLOORS

Knowledge of water depths in shallow seas has obviously always been important for sailors, and bathymetric charts for coastal waters have been produced for several centuries. The need for more extensive information about ocean bathymetry did not arise until relatively recently, and the first ocean-wide bathymetric chart was published by Matthew Maury in 1855, for the North Atlantic (Figure 1.10). Maury's 'Middle Ground' was the first indication of the existence of the Mid-Atlantic Ridge, the long and sinuous form of which was soon revealed by the *Challenger* soundings, despite the limited number of measurements that could be made (Section 1.1.2).

The first contoured charts to cover the whole of the world's ocean floor were published in Monaco in 1903, in 24 sheets at a scale of 1 : 10 million, as the first edition of the General Bathymetric Chart of the Oceans (GEBCO). The fifth edition of the GEBCO charts was published in 18 sheets between 1975 and 1982, with contours at 500 m intervals in the deep oceans, under the guidance of the Intergovernmental Oceanographic Commission and the International Hydrographic Office. This edition is now digitized to facilitate continual updating, and is available on CD-ROM.

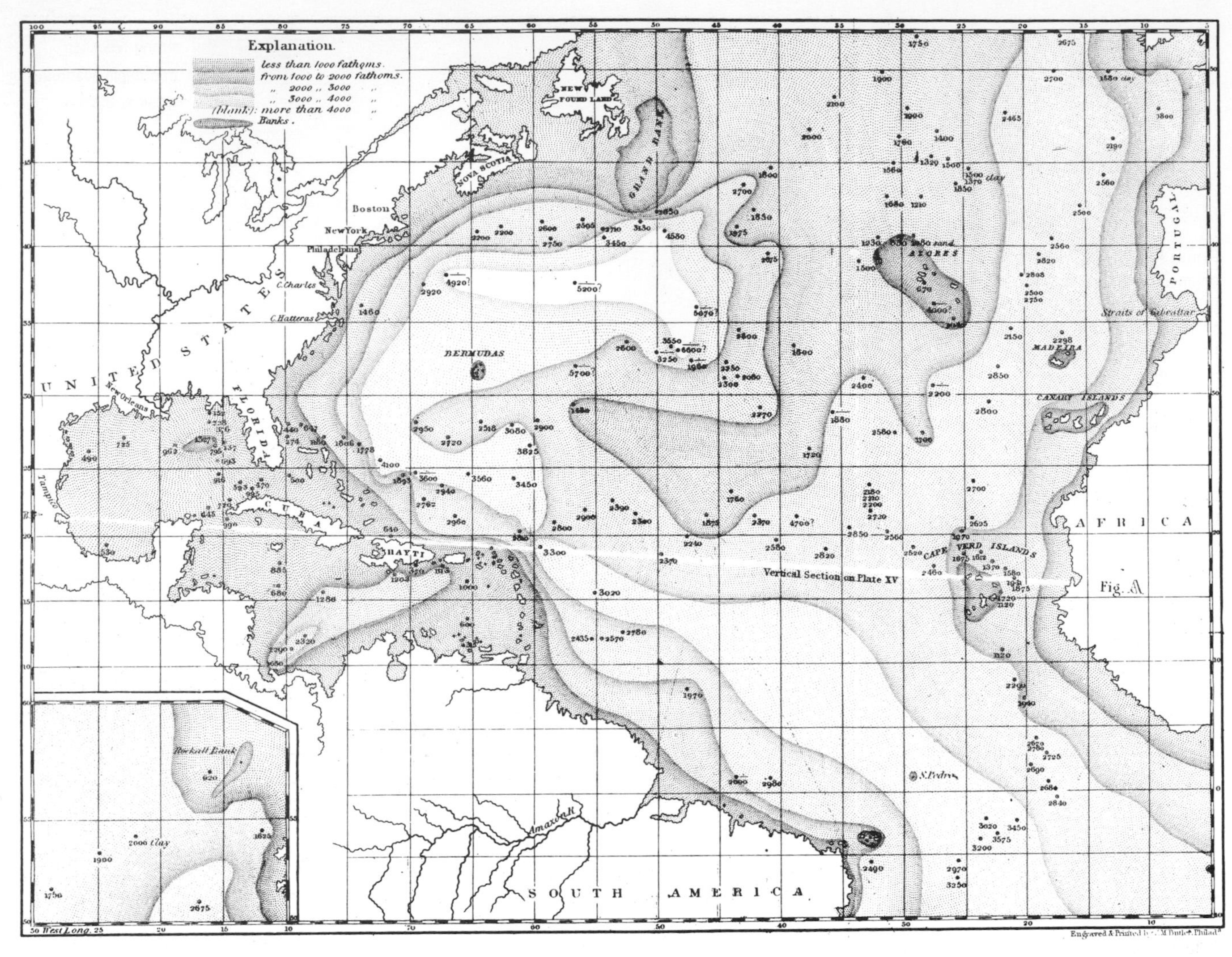

Figure 1.10 (a) Map of North Atlantic bathymetry by Matthew Maury, 1855.
Source: M. Maury (1857) *The Physical Geography of the Sea*, Sampson Low & Son.

The physiography of the ocean floor is now thought to be quite well known, but until recently even some large features had not been recognized, and much detail still needs to be added. Thus, it was not until the 1960s that the International Indian Ocean Expedition delineated the Ninety-east Ridge (Figure 1.11) as a single 4500 km-long feature. Submerged volcanoes (seamounts) and other features are still being discovered in the oceans by satellites using altimeter measurements of the sea-surface (see Section 1.2.1).

The need for more extensive and precise bathymetric data continues to grow, for practical as well as academic reasons. Detailed charts of continental shelf and slope bathymetry are essential for exploration and exploitation of sea-bed mineral resources (e.g. petroleum, sands and gravels), and such areas may need to be resurveyed frequently if sediment movement is rapid. Bathymetric data can also aid navigation, because echo-sounding can detect features such as submarine canyons, which in well-charted waters are sufficiently defined to permit accurate position-fixing. In the open oceans, bathymetric information is essential for studies of mid-ocean ridges, deep water circulation, and the movement of sediments.

Echo-sounding is a great advance on the early ‘spot’ soundings with lead lines, but what is the limitation of this method for ocean-floor mapping?

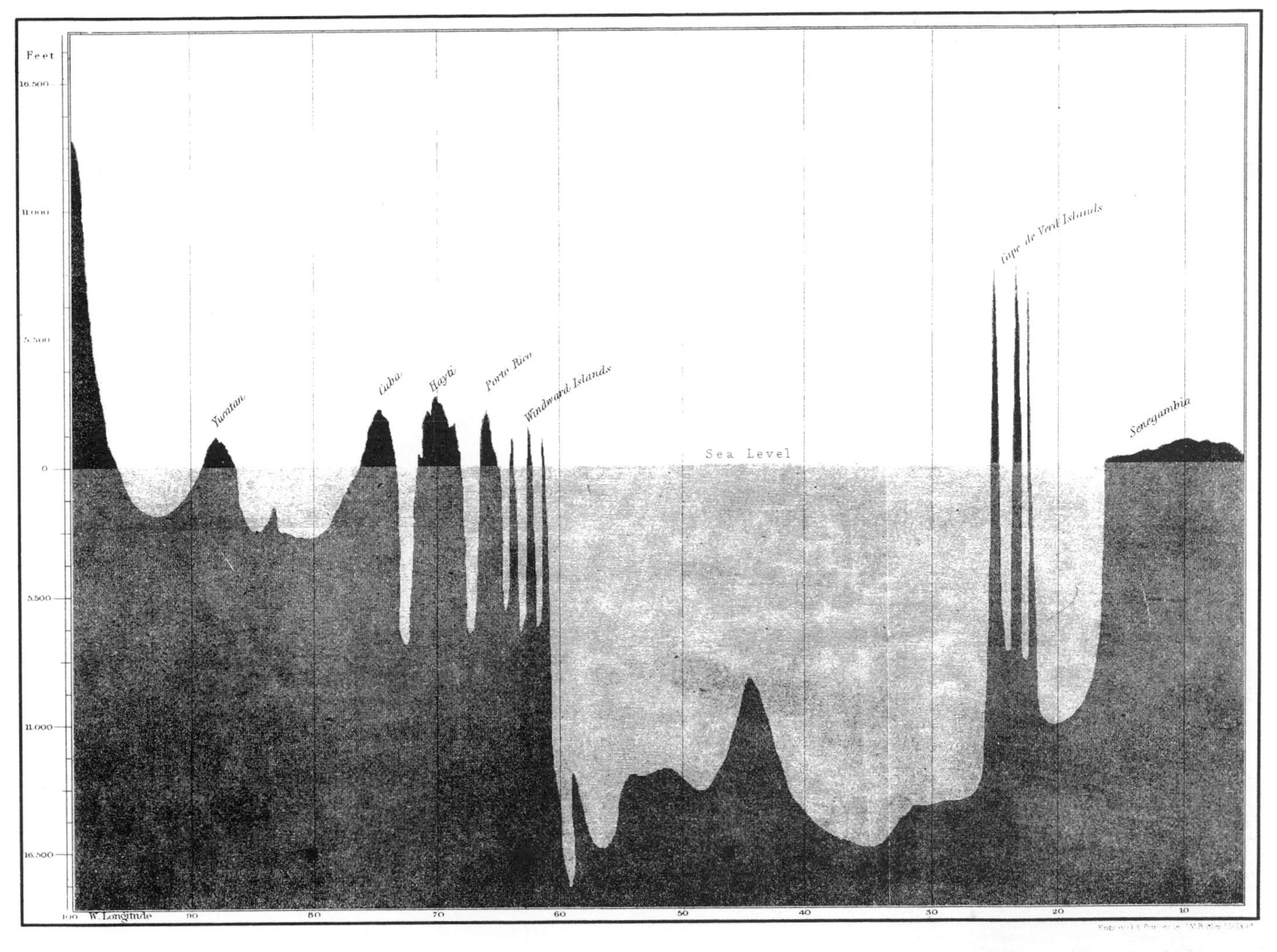

Figure 1.10 (b) Profile of North Atlantic bathymetry, along the white line across Figure 1.10(a), by Matthew Maury, *op. cit.* Note that the vertical exaggeration is very great (about × 500).

Echo-sounder records provide linear bathymetric profiles along the ship's track. A large amount of interpolation is necessary between profiles unless there are plenty of them along tracks that intersect one another at frequent intervals.

For detailed survey work, 'swath bathymetry' data can be obtained by using a fan-like array of many narrow sonar beams directed to either side of the towing ship's track (Figure 1.12(a), p.17). This allows bathymetry to be mapped along several parallel lines at once, provided the longer path length of the non-vertical beams is allowed for. One of the earliest and best known swath bathymetry echo-sounders is *Sea Beam*, which came into service in 1977. This emits a fan of 16 narrow (2.7°) beams, spreading out to cover a swath whose width is about two-thirds of the water depth. More modern multibeam swath bathymetry systems use a hundred or so narrow beams and can cover swaths whose width is up to four times the depth of water between the instrument and the sea-bed. However, this does not eliminate the need for interpolation into unsurveyed areas, and there are vast areas of ocean floor still relatively poorly mapped by any kind of echo-sounding.

Since the 1960s, side-scan sonar systems have been developed. These differ from swath bathymetry systems in that they scan the sea-floor with two divergent sonar beams – as distinct from 'fans' of narrow beams – that acoustically illuminate broad strips of sea-floor on either side of the ship's

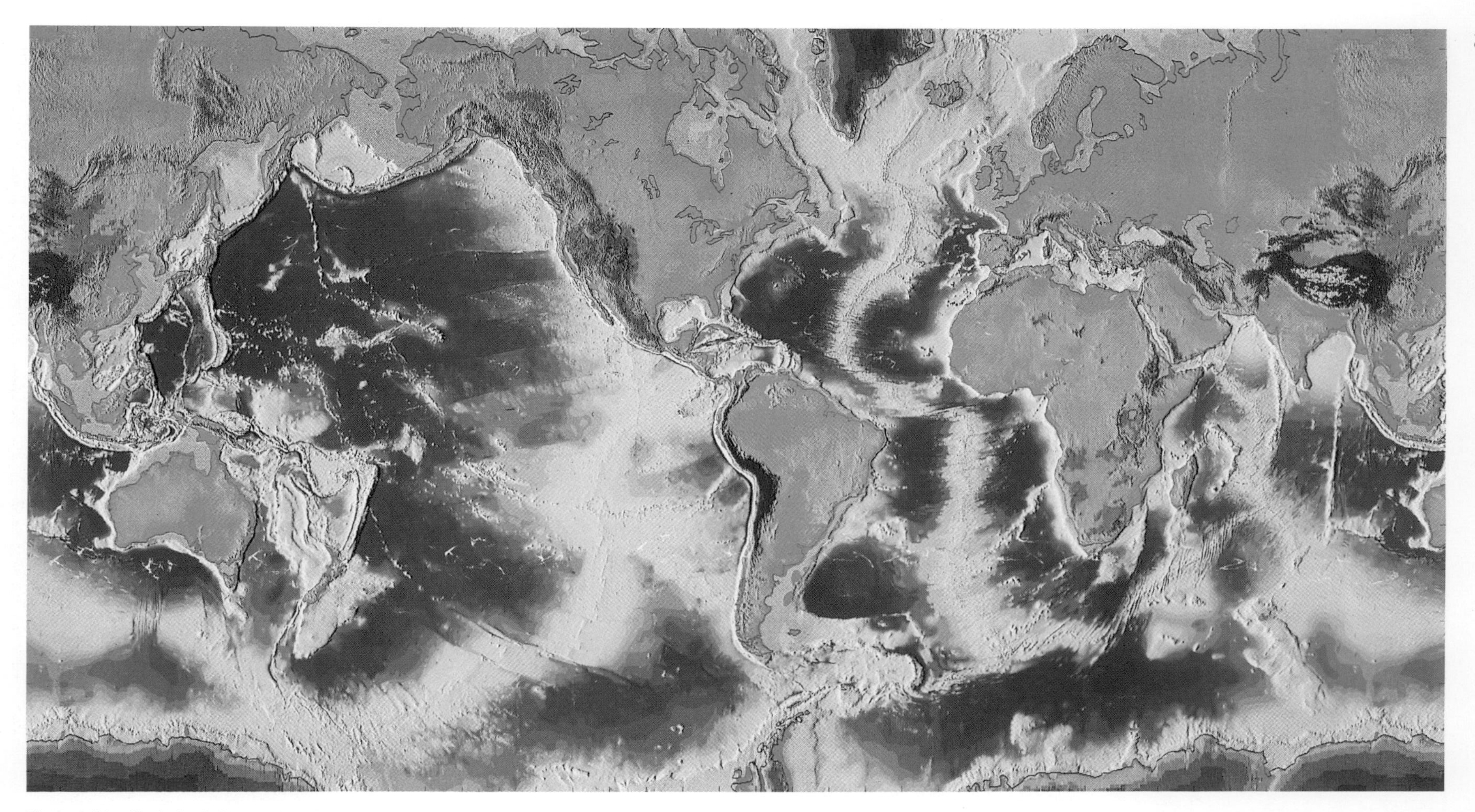

Figure 1.11 Shaded relief map of the Earth's solid surface. The Ninety-east Ridge is the linear ridge running almost north–south to the south-east of India near the right-hand edge of the map.

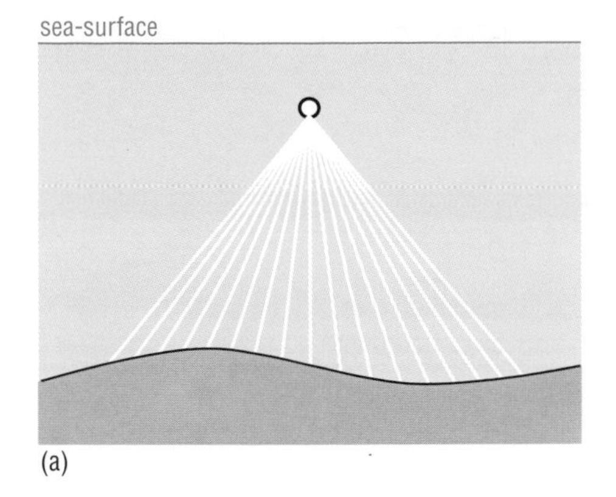

(a)

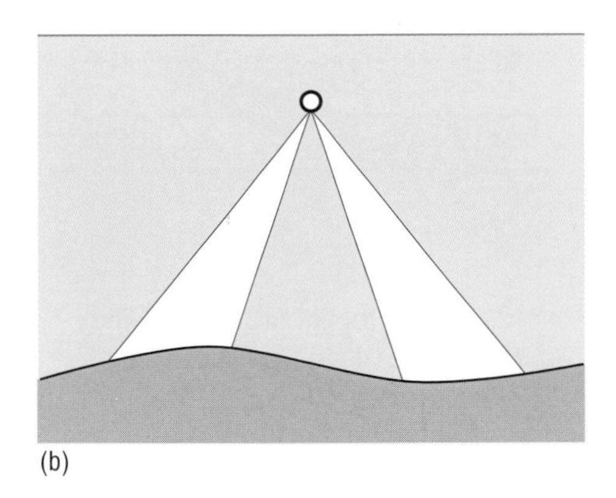
(b)

Figure 1.12 (a) Swath bathymetry.
(b) Side-scan sonar.

track (Figure 1.12(b)). Complex processing of the echoes enables *images* of the sea-floor to be produced, rather than actual maps of bathymetry. Like swath bathymetry systems, side-scan sonar systems are typically housed in torpedo-like 'towfish' casings towed behind the ship. Some are designed to be towed just below thc sca-surface, enabling a survey of wide areas. The best known examples are *GLORIA* (Figure 1.13(a)) and *SeaMARC II*. Other side-scan sonars such as the one mounted in *TOBI* (Figure 1.13(b)) are designed to be towed only a few hundred metres above the sea-floor. Such deep-tow devices can image individual objects only a few metres across, but take much longer to survey a given area because of the narrowness of their imaging swath and their slower towing speed.

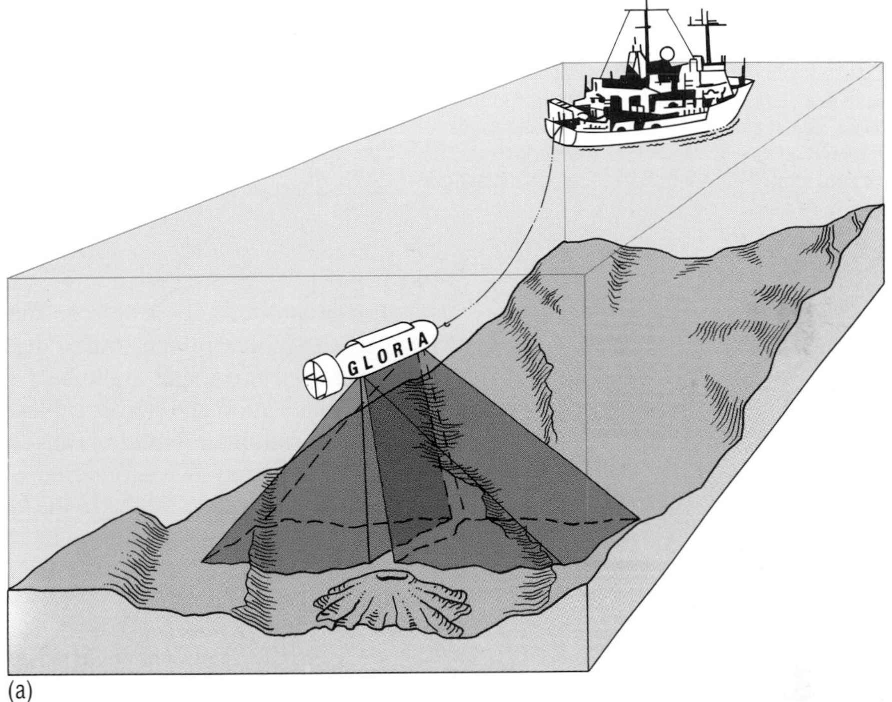

(a)

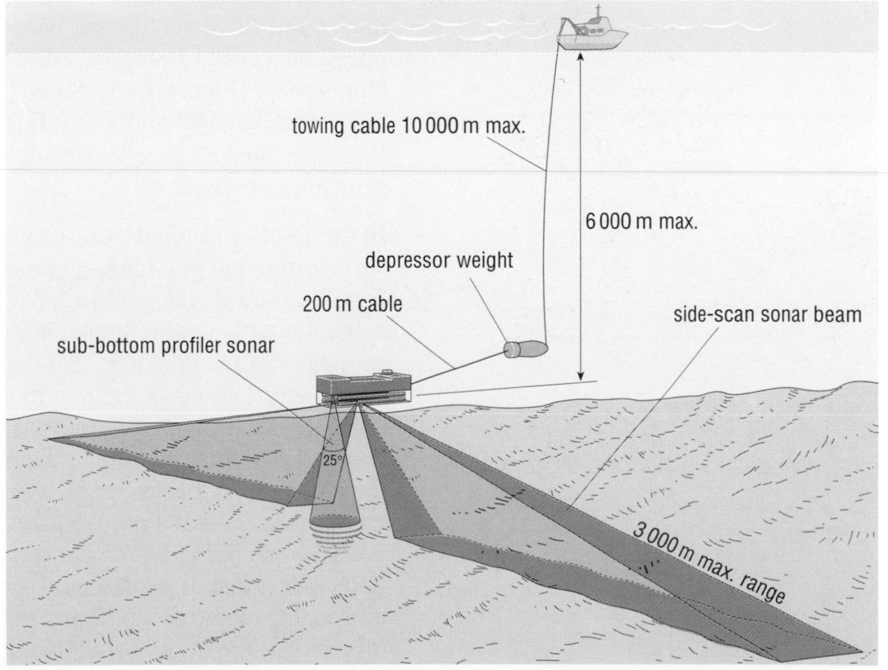

(b)

Figure 1.13 (a) *GLORIA* (*G*eological *LO*ng *R*ange *I*nclined *A*sdic) is 8 m long and is towed 300 m behind the ship at a speed of 10 knots and a depth of 50 m, where it is neutrally buoyant – i.e. the overall density of the 'towfish' package is the same as that of the surrounding seawater. In water of 5000 m depth, *GLORIA* scans two 30 km-wide swaths of sea-floor to either side of its track; the time interval between sonar pulses (of 4 s duration in the 6.2–6.8 kHz range) is set at 40 s to allow time for the most distant echoes to return. In shallower water, the area covered and the time between pulses are reduced. *SeaMARC* is similar but uses narrower beams of a higher frequency (*c.* 30 kHz).
(b) *TOBI* (*T*owed *O*cean *B*ottom *I*nstrument) is a 4.5 m-long deep-tow device, which is towed at a speed of about 2 knots about 400 m above the sea-floor. As well as a 30 kHz side-scan sonar, *TOBI* carries a 7.5 kHz sub-bottom profiler sonar, a magnetometer, a temperature probe, and a transmissometer to measure transmission of light along a 25 cm path.

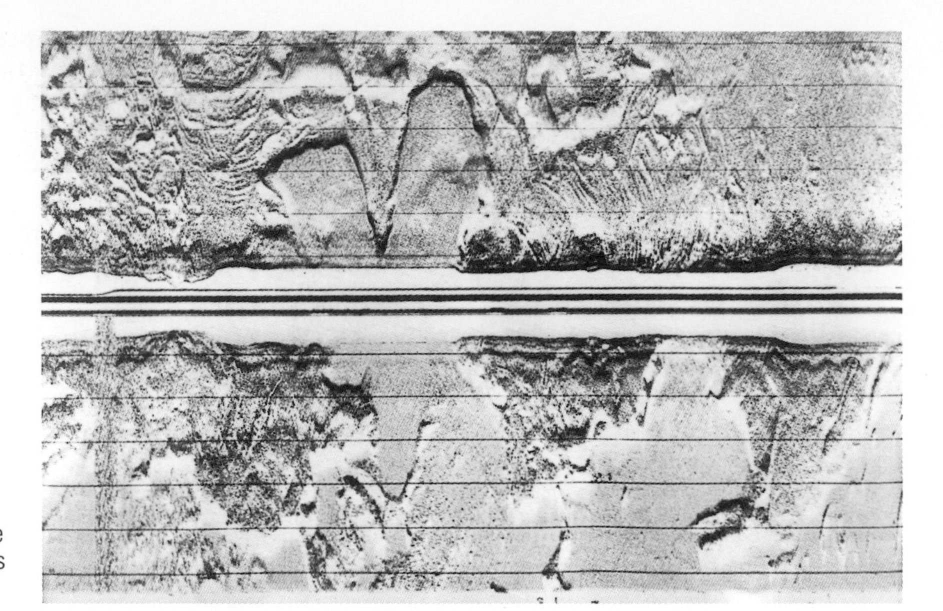

Figure 1.14 Part of an image obtained from a side-scan sonar traverse at the entrance to the Black Sea. It shows bare rocky areas and sand-covered areas with sandwaves (megaripples). The central strip is the ship's path. The horizontal lines are 15 m apart.

Images produced by side-scan sonar not only reveal local bathymetry, they can also indicate the nature of the bottom (Figure 1.14), information that is essential if breakages are to be avoided when laying pipelines or communication and power cables. Features which have been revealed include sinuous submarine canyons, submarine slides, thousands of previously unknown submarine volcanoes, metal-rich deposits and manganese nodule fields. In the late 1980s, deep-towed side-scan sonar imaging systems began to be modified to record echoes in a complex fashion, allowing highly detailed swath bathymetry to be extracted from the data as well as images.

1.2.1 BATHYMETRY FROM SATELLITES

In spite of the many depth measurements made since the 1920s, especially during and since the Second World War, some parts of the oceans remain poorly surveyed by ship-based survey, especially in the Southern Hemisphere (Figure 1.15). Since the 1970s, measurements from satellites have helped considerably to refine available bathymetric maps of the oceans. We shall consider specific examples in Chapter 2, but here we focus on the principles involved.

In the absence of winds, tides and currents, the sea-surface would follow a surface of equal gravitational potential known as the **geoid**. Because the Earth is rotating, the geoid approximates to a slightly flattened sphere (a spheroid) with a polar diameter about 43 km less than its equatorial diameter. Variations in the thickness and density of the Earth's crust and upper mantle cause perturbations in the gravity field, so the geoid is not actually a smooth spheroid; it has long (hundreds to thousands of km) wavelength highs and lows, with a relief relative to the reference spheroid of up to 200 m across the entire globe. Many people find the concept of the geoid confusing. The crucial point is that although it has 'highs' and 'lows', it is a surface of equal gravitational potential. This means that if we could cover the geoid surface with balls, none of them would show any tendency to roll 'downhill'. Also, a plumb-line would always hang at right angles to the surface of the geoid, so it is the surface of the geoid that defines what we think of as 'horizontal'.

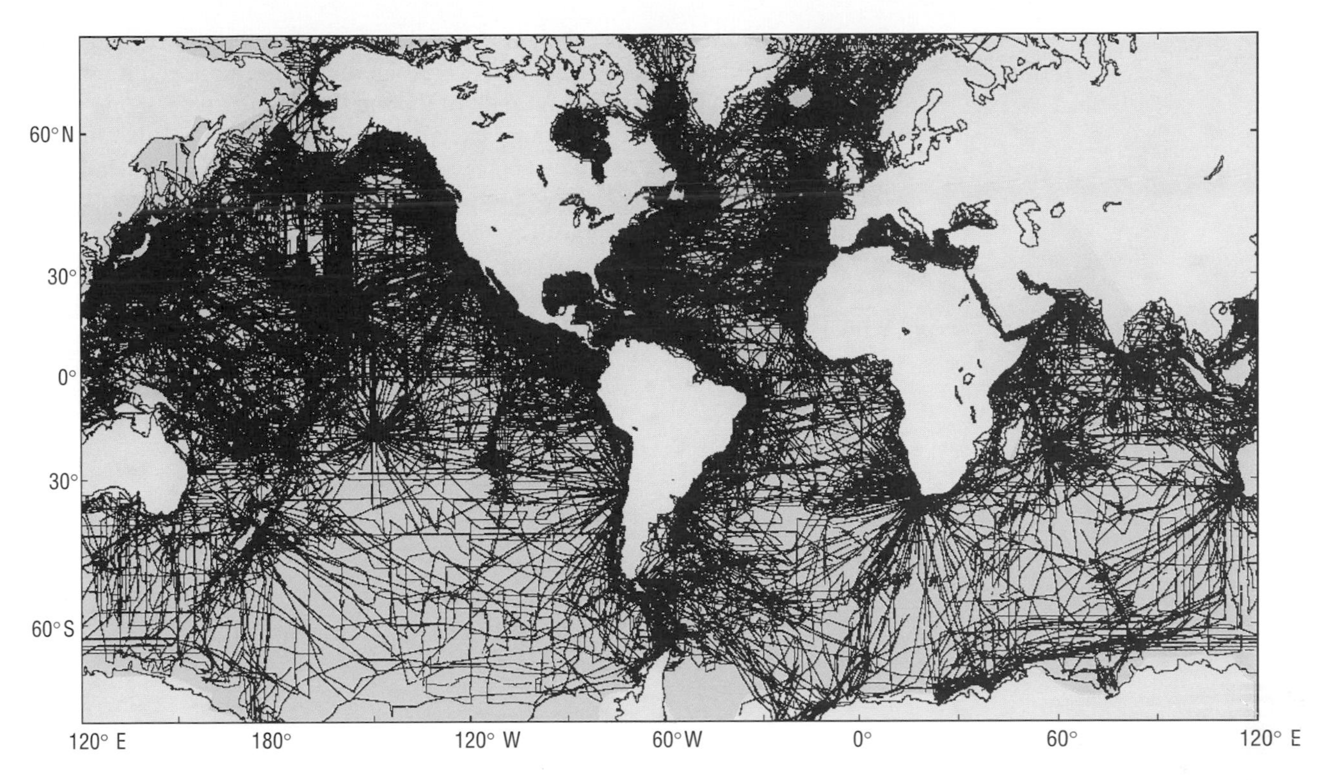

Figure 1.15 Bathymetric survey lines available for compilation of GEBCO charts and collected by ships up to October 1995. Much of the data collected along these lines are old and of dubious quality. At the present rate, it would take about 125 years to survey the oceans completely using modern swath bathymetry techniques.

The significance of the geoid for bathymetry is that shorter wavelength, lower amplitude relief corresponding to the topography of the sea-floor is superimposed on the long wavelength relief caused by variations within the crust and upper mantle. For example, the extra mass of a 2 km-high seamount attracts extra water over it, causing a 'bulge' in the sea-surface about 2 m high and about 40 km across. Similarly, the reduced gravity over trenches in the sea-floor means that less water is held over these regions, so that the sea-surface is locally depressed. This means that a technique that can map the height of the sea-surface can also be used to pick out bathymetric features (Figure 1.16).

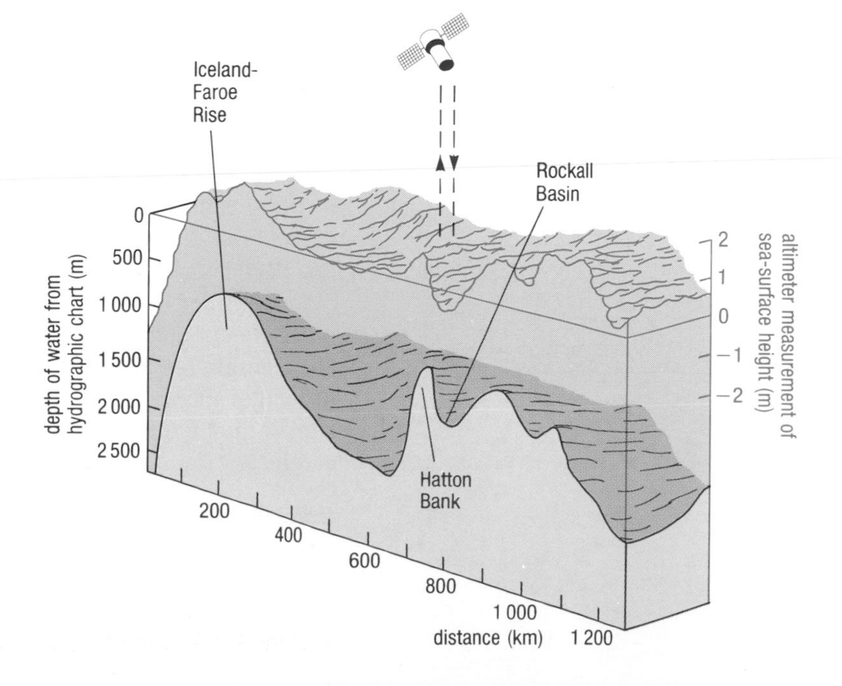

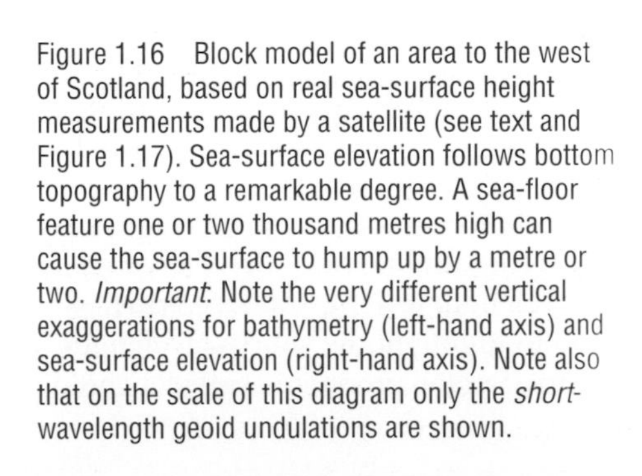

Figure 1.16 Block model of an area to the west of Scotland, based on real sea-surface height measurements made by a satellite (see text and Figure 1.17). Sea-surface elevation follows bottom topography to a remarkable degree. A sea-floor feature one or two thousand metres high can cause the sea-surface to hump up by a metre or two. *Important*: Note the very different vertical exaggerations for bathymetry (left-hand axis) and sea-surface elevation (right-hand axis). Note also that on the scale of this diagram only the *short*-wavelength geoid undulations are shown.

Satellites that measure their height above the sea-surface provide exactly the sort of data required to do this, and several satellites have achieved it by means of radar altimetry. The altitude of the satellite is measured as in narrow-beam sonar echo-sounding, except that a series of radar pulses is used instead of sound, and the echoes come from the sea-surface instead of the sea-floor (Figure 1.17). The 'footprint' of the radar beam is several kilometres across, so irregularities in surface height caused by wind-generated surface waves are averaged out. To derive the shape of the geoid, the satellite's position at all times must be known with equal precision, which can be achieved by means of lasers and Doppler tracking.

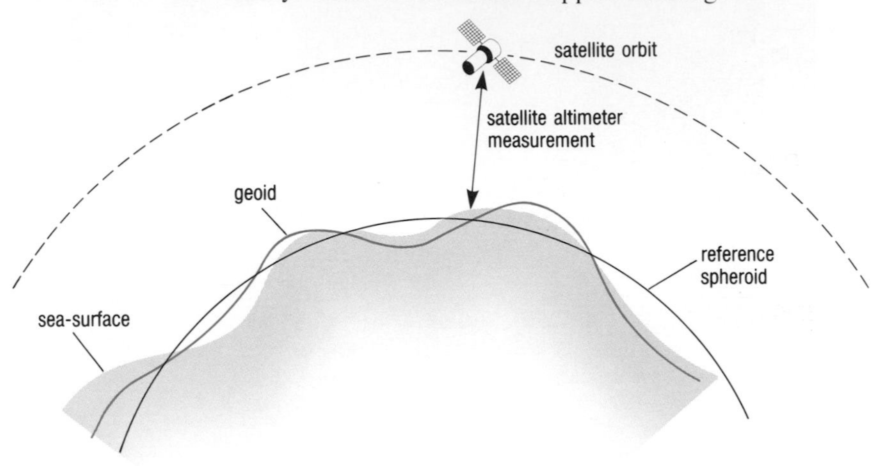

Figure 1.17 A satellite radar altimeter measures its own height above the sea-surface. To reveal smaller-scale bathymetric features in particular, there is an important correction to be made. Tides, currents and changes in atmospheric pressure can cause undulations of more than a metre in the ocean surface *above* or *below* the surface of the geoid. These variable influences must be averaged out if bathymetric details such as those depicted in Figure 1.16 are to be detected. Note that scales on this diagram are differently exaggerated: geoid undulations (relative to a smooth reference spheroid) are two orders of magnitude greater than sea-surface topography.

QUESTION 1.2 Can you suggest a way of allowing for the dynamic and therefore variable perturbations of the sea-surface referred to in the caption to Figure 1.17, when attempting to make corrections to bathymetric measurements obtained from satellite data?

The first global radar altimetry mapping was achieved by the *Seasat* satellite in 1978, and even though the altimeter failed after only three months *Seasat*-derived marine geoid data were a mainstay of oceanographers for nearly two decades. A US Navy satellite *Geosat* launched in 1985 had more closely spaced repeating ground paths than *Seasat*, and was tracked precisely enough to map the geoid at a horizontal resolution of 10–15 km and a height resolution of 3 cm. Most of these data remained classified until the end of the Cold War, but were released to complement altimetric data of similar quality obtained by the European satellites *ERS*-1 (launched 1991) and *ERS*-2 (launched 1995).

To be of use for mapping sea-bed topography, altimetrically-derived geoid maps are filtered to suppress the long-wavelength relief associated with crustal and mantle density variations, and to highlight the short-wavelength anomalies that correspond with bathymetry. Such a product is called a geoid anomaly map (Figure 1.18).

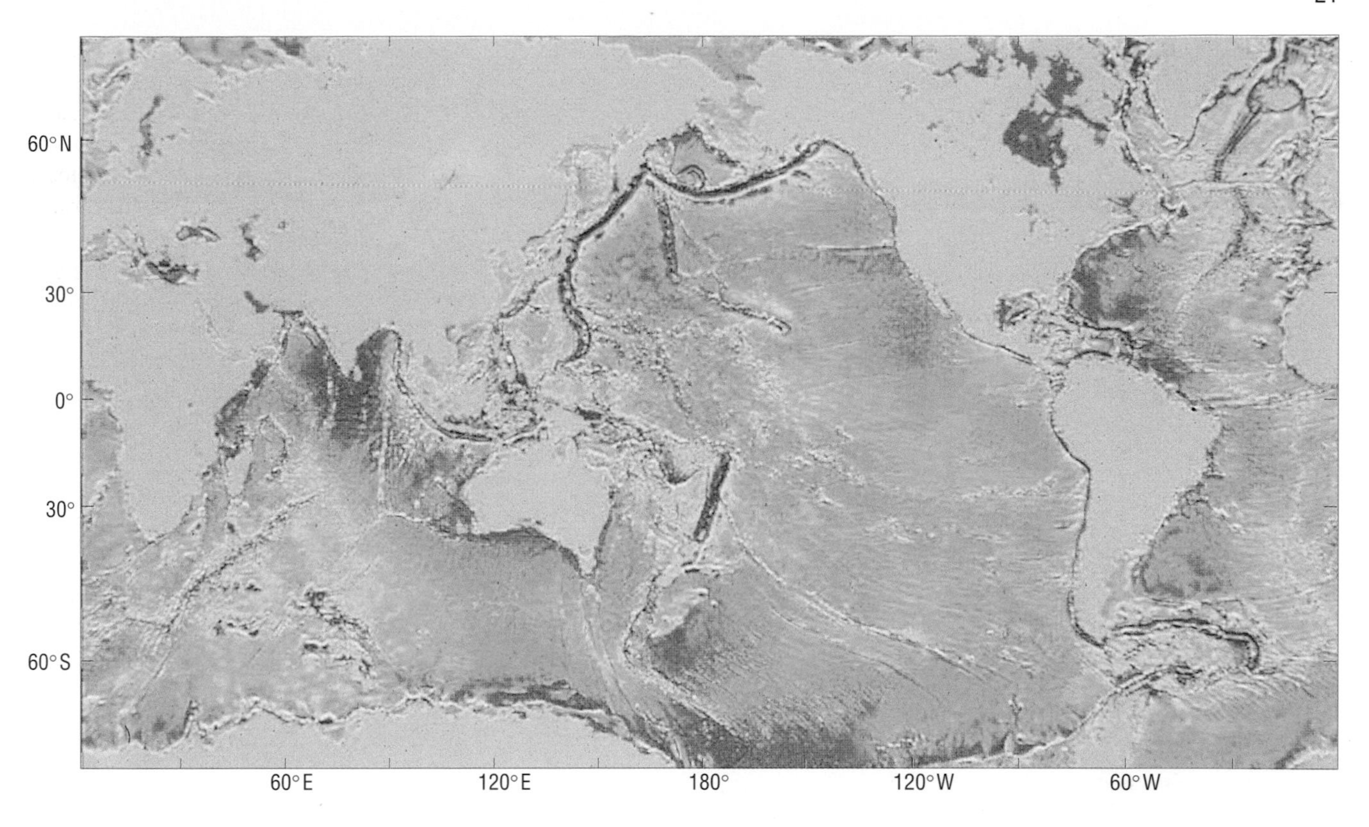

Figure 1.18 Global geoid anomaly map derived from four-and-a-half years of *Geosat* and two years of *ERS*-1 radar altimeter data. The data have been filtered to show detail but reduce the effect of very large-scale features. In green areas, the sea-surface lies close to the mean geoid. Yellow–orange–red hues represent increasing height above the mean geoid (corresponding to stronger gravity over positive relief features on the sea-floor), and blue–violet–magenta hues represent increasing depression below the mean geoid (corresponding to weaker gravity over negative relief features on the sea-floor). The spatial resolution is limited by the effective size of the radar 'footprint', which is a few km across.

Another remarkable outcome of the space programme has been the discovery that radar data from satellites can reveal the configuration of the sea-floor in shallow water. In this case, the instrument is not an altimeter, but an imaging radar, which constructs an image of the sea (or land) surface from a complex series of radar echoes obtained from one side of the satellite's track. This technique (known as synthetic aperture radar, or SAR) is analogous to side-scan sonar, and produces images such as that shown in Figure 1.19. However, the radar beam is incapable of penetrating more than about 1 cm below the sea-surface, so it is clear that the sand-banks and other bottom features are not being 'imaged' directly. What happens is that tidal currents flowing over an irregular sea-floor create a pattern of sea-surface roughness which is related to the bottom topography. These patterns are not in themselves a new discovery; indeed, they appear to have been known to the English captains blockading the Spanish Armada in French ports in the sixteenth century. The real discovery was the close correlation between bottom features and sea-surface ripples of a wavelength that could interact with that of the radar beam (23.5 cm). Surveys of the sea-bed which otherwise could take weeks or months from a surface vessel can be achieved in a few seconds, if the satellite makes its pass at the right phase of the tidal cycle.

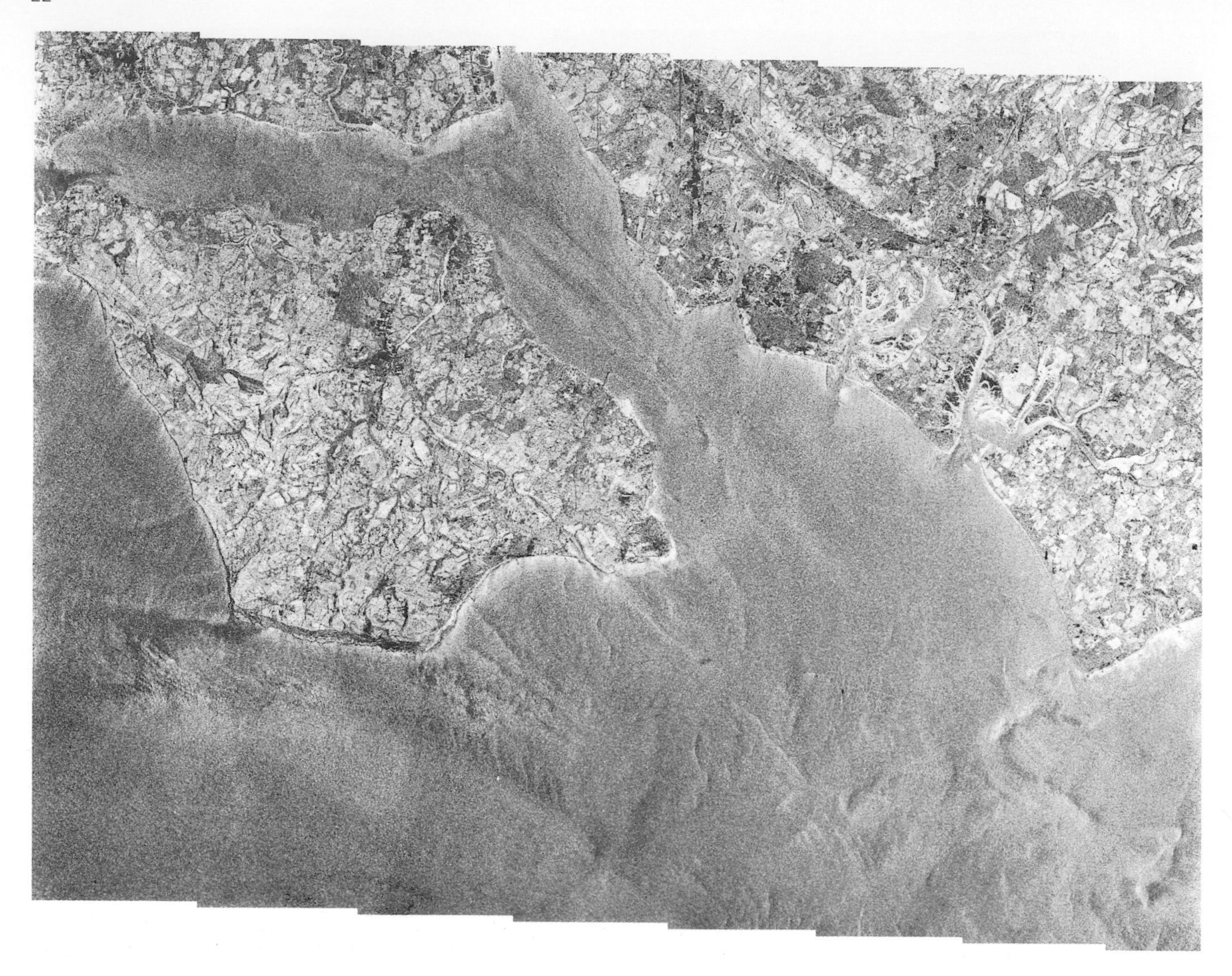

Figure 1.19 Patterns of sand-banks with sandwaves (megaripples) superimposed on them, and other bathymetric features, show up in this *Seasat* SAR image of part of the English Channel.

1.3 UNDERWATER GEOLOGY

Before 1930, virtually the only data on the nature and composition of the oceanic crust were from samples of rock and sediment recovered embedded in wax or tallow fixed to the base of sounding-line weights or else obtained by dredging, by *Challenger* and later expeditions. Geophysical investigations in the decades immediately before and after the Second World War showed that the oceanic crust is denser than continental crust and has a distinctly layered structure. At the time, the consensus among geologists was that the oceans and continents were fixed in their positions. Subsequent studies became caught up in the plate tectonics 'revolution' of the 1960s and early 1970s, which revealed just how fundamentally the oceans differ from the continents.

The Deep Sea Drilling Project (DSDP) was begun in the 1960s using the *Glomar Challenger* (Figure 1.20), a ship that was specially equipped for drilling into the ocean floor beneath several kilometres of seawater. The main technical advance which made this possible was the development of a re-entry system whereby the bits could be replaced and the drill string repositioned in the hole (Figure 1.21). Many hundreds of holes have now been drilled through the ocean-floor sediments into the underlying igneous

Figure 1.20 *Glomar Challenger*. Thrusters in the bow and stern were controlled by computer to maintain station within 30 m of a point directly above an acoustic beacon on the ocean floor. Drilling could thus proceed without snapping the drill string.

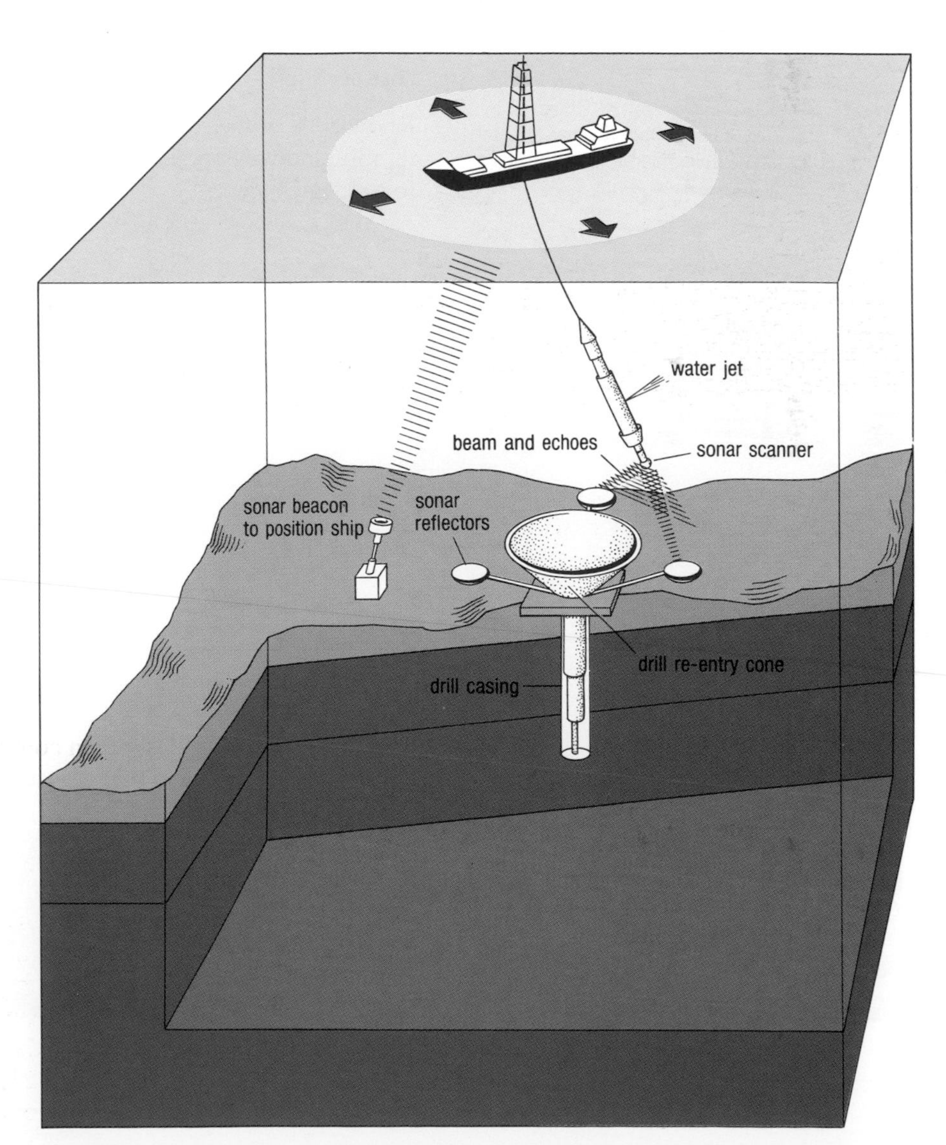

Figure 1.21 (Not to scale). Re-entering a drill hole in the ocean floor: the re-entry cone is attached to the drill 'string' when it is first lowered to the bottom and it remains there when the string is withdrawn for the drill bit to be replaced. For re-entry, the drill string is lowered again, with a sonar scanner mounted on the bit assembly. This emits sound signals that are echoed back from three reflectors spaced round the cone. Information about the position of the bit assembly relative to these reflectors is used in the control of water jets which steer the bit directly over the cone.

rocks, and have added greatly to our understanding of crustal structure and sediment distribution. The *Glomar Challenger* was succeeded by a similar but better-equipped vessel, the *JOIDES Resolution*, which began operations in the international Ocean Drilling Program (ODP) in 1985.

Examination of the oceanic crust has been further facilitated by the use of instruments lowered down drill holes to measure the physical and chemical properties of the sediments and underlying volcanic rocks *in situ* (see also Section 4.2.5). Drilling has been supplemented by increasingly sophisticated echo-sounding techniques, side-scan sonar and underwater photography from cameras towed close to the sea-floor, which have enabled very accurate bathymetric maps and profiles to be compiled.

Since 1963, geologists have been able to do 'fieldwork' under several kilometres of seawater using manned submersibles (Figure 1.22), equipped with precise navigation systems (including sonar), sampling apparatus and cameras. The use of remotely controlled submersibles achieved great publicity in 1986 when one was used to take video pictures inside the wreck of the *Titanic*. Nowadays, such vehicles typically carry cameras, side-scan sonar equipment, and sampling gear, and can reduce the time taken for deep-sea exploration and survey from years to weeks, because they can dive deeper and stay down longer than manned submersibles. They are also cheaper and involve virtually no risk to human life.

Knowledge of the form, structure and evolution of ocean basins is growing rapidly, and we hope to give you some flavour of these developments in the following Chapters.

Figure 1.22 The submersible *Alvin* at the start of a dive. *Alvin* has been used to investigate many key areas of the ocean floor.

1.4 SUMMARY OF CHAPTER 1

1 Accurate navigation at sea became possible only with the development of accurate and reliable chronometers in the eighteenth century, which enabled longitude to be determined. Nowadays, there are many radio-navigational aids, both surface and satellite-mounted.

2 Accurate depth determinations became possible only with the development of echo-sounding early in this century. For detailed survey work, swath bathymetry is now favoured, using multiple narrow beams directed to either side of the ship's track.

3 Direct imaging of the sea-floor, as distinct from bathymetric mapping, is achieved by side-scan sonar instruments. These image strips of sea-floor as much as about 10 km wide from a single ship track. Deep-tow devices allow narrower swaths to be imaged at higher resolution.

4 Satellite altimetry uses radar to measure very precisely the average height of the sea-surface, which follows a surface of equal gravitational potential – the geoid. After transient effects (tides, currents and atmospheric pressure changes) have been averaged out, the mean height of the sea-surface can be determined. Short-wavelength irregularities in the geoid (geoid anomalies) correspond to topographic features on the ocean floor, and the height of the sea-surface can therefore provide a measure of bathymetry. Imaging radar (SAR) can show up bathymetric features in shallow water.

5 Detailed exploration of oceanic crustal structure and composition began in the 1960s, with the DSDP, succeeded in 1985 by the ODP. Much has been learned with the help of new technologies for imaging and through submersible operations.

Now try the following question to consolidate your understanding of this Chapter.

QUESTION 1.3 Estimate the approximate wavelength and amplitude of variations in the geoid in the area depicted in Figure 1.16.

CHAPTER 2 THE SHAPE OF OCEAN BASINS

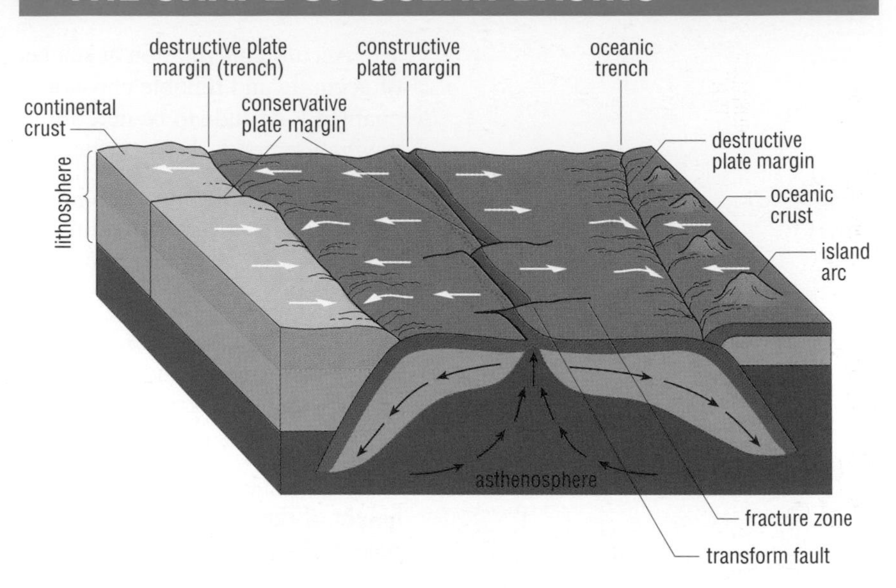

Figure 2.1 Vertically exaggerated diagram showing the basic concepts of plate tectonics. Plates of rigid lithosphere (comprising oceanic and/or continental crust and uppermost mantle) mostly about 40–60 km thick in oceanic regions and about 100–200 km thick in continental regions overlie a layer of relatively low strength in the mantle called the asthenosphere. (The crust/mantle distinction is based on composition, the lithosphere/asthenosphere distinction is based on physical properties.) Mantle material rises from the asthenosphere at constructive plate margins (also known as ocean ridges or spreading axes) and is added to the plates. Old plate material descends into the asthenosphere at destructive plate margins (ocean trenches). At conservative plate margins, plates merely slide past each other. (Mature island arcs, discussed later, are best regarded as continental rather than oceanic crust.)

Few students of Earth sciences have not seen maps and diagrams such as Figures 2.1 and 2.2, or heard of **sea-floor spreading** and **plate tectonics**. We provide below a numbered summary of the main features relating to these processes, partly as a reminder and partly as a basis for what follows in this Volume.

1 The **mantle** forms the majority of the Earth's volume, extending right down to the core. It has the composition of the rock type known as **peridotite**. Above the mantle is the **crust**, which differs from it chemically.

2 The crust and the mantle immediately beneath it together form a single rigid layer known as the **lithosphere**, which comprises a number of **tectonic plates**. The mantle below the lithosphere is solid, but it is not rigid and is convecting at speeds of the order of centimetres per year (this is how heat is transported from the Earth's interior to the base of the lithosphere). The upper part of this convecting region of the mantle is particularly weak, and is called the **asthenosphere**. It is this weak layer that enables the plates to move.

3 The lithosphere is up to 200 km thick under the continents but usually only about 40–60 km thick under the oceans*.

4 The Earth has two types of crust: **continental crust** (underlying virtually all land areas and shallow seas), and **oceanic crust** (the floors of the oceans). Most plates have areas of both oceanic and continental crust, but the Pacific Plate (Figure 2.2) has relatively little continental crust, and some plates have *only* oceanic crust (e.g. Nazca Plate, Figure 2.2).

5 Continental crust has an average thickness of 35 km, but is up to 90 km thick beneath mountain ranges. In composition, it is mostly granitic or andesitic (i.e. intermediate between granite and basalt).

6 Oceanic crust is both thinner (7–8 km) and denser than continental crust. It is basaltic in composition. Note that here we are not including sediments as part of the oceanic crust. Sometimes, oceanic crust is used to mean the igneous crust *and* the overlying sediments.

* If you have studied geophysical aspects of the Earth sciences, you may realize that the lithosphere referred to here is the plate tectonic lithosphere (roughly equivalent to the elastic lithosphere), rather than the thermal lithosphere, which is somewhat thicker. In this Volume, there is no need to worry about such subtleties, and the definition given above (lithosphere = the Earth's outer, rigid shell) is perfectly adequate.

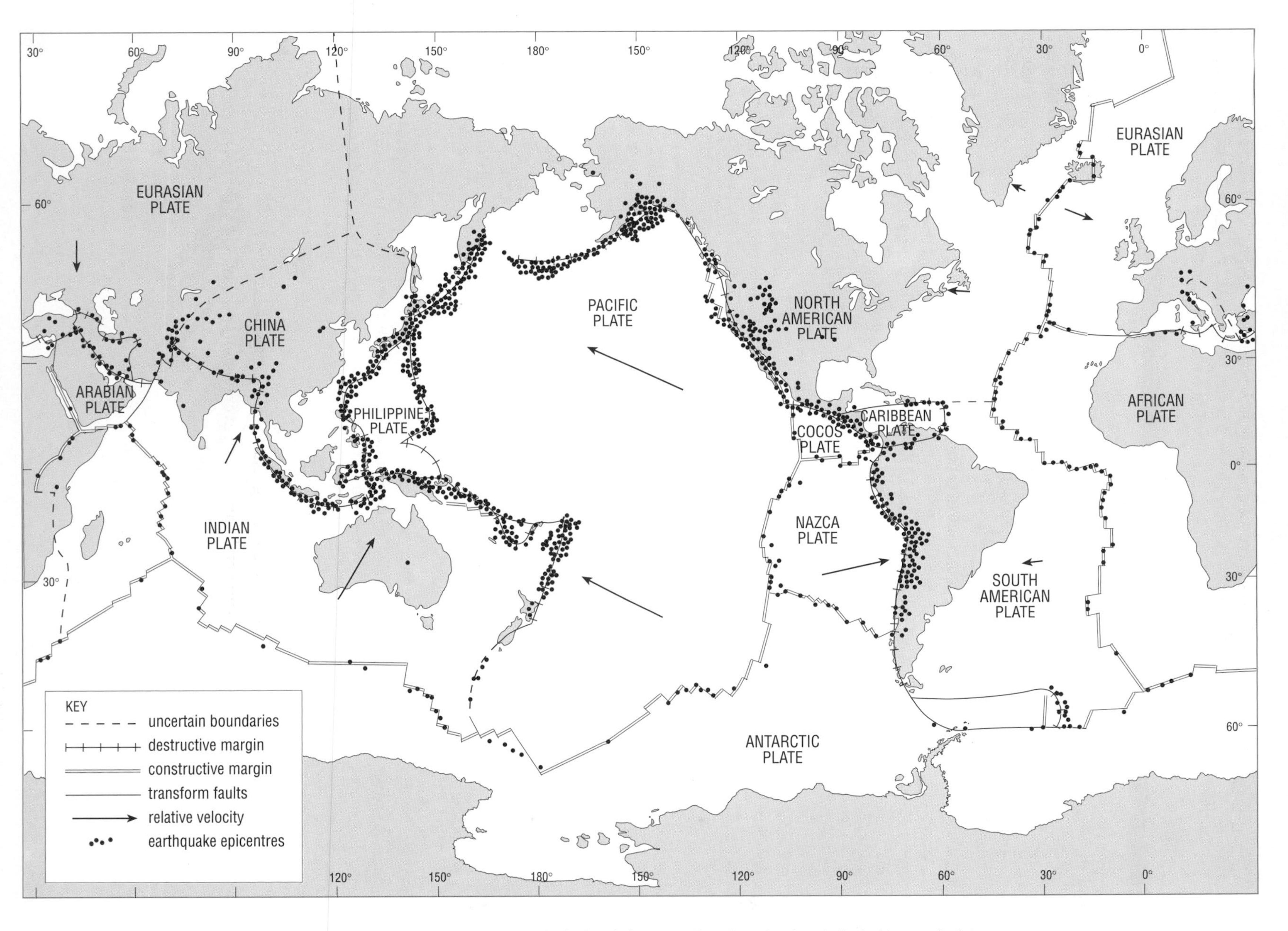

Figure 2.2 The world pattern of plates, ocean ridges, trenches and transform faults in relation to earthquake epicentres indicated by purple dots. Tentative positions of plate margins are indicated by dashed lines. There are seven major plates and six minor ones, plus several smaller ones not named here. The length and direction of the arrows indicate the relative velocities of the plates, averaged over the past few millions of years (Ma). The African Plate is assumed to be stationary. The arrow length in the key corresponds to a relative velocity of 5 cm yr^{-1}.

7 Oceanic crust lies well below sea-level whereas continental crust is mostly much higher. This is a result of **isostasy**: the Earth's outer layer tends towards gravitational equilibrium by height adjustments between areas of different relative buoyancy.

What topographic features on Figure 1.11 correspond to the **constructive** and **destructive plate margins** illustrated in Figures 2.1 and 2.2?

They are described below in items 8 and 9.

8 The global system of constructive margins or sea-floor **spreading axes** forms the nearly continuous ridge system that resembles a mountain chain snaking its way through the major ocean basins. This is where oceanic lithosphere is generated.

9 The numerous ocean trenches and island arcs – most of them within the Pacific – are sites of oceanic plate **subduction** and re-absorption into the asthenospheric mantle. The Alpine–Himalayan mountain chain bears witness to a phase of subduction and continental plate collision that is geologically quite recent (within the past 150 Ma).

10 The polarity of the Earth's magnetic field has reversed at intervals of a few hundred thousand to a few million years (Ma), during the past 100 Ma or so (before that, the intervals appear more variable). Alternating 'normally' magnetized and 'reversely' magnetized stripes of ocean floor lie parallel to the constructive margins at which they were formed (Figure 2.3). These enable the age of the ocean crust to be mapped and sea-floor spreading histories to be compared among the different ocean basins.

11 No oceanic crust older than about 180 Ma has been found in any of the major oceans, so the present-day ocean basins are very young in relation to the age of the Earth (4600 Ma). Furthermore, they are changing all the time, in two major respects:

(a) Shapes and sizes are changing because of plate movements. The Atlantic and Indian Oceans are getting bigger, because they have spreading axes but no major subduction zones. In the Pacific, subduction outpaces spreading so this ocean is contracting. The rates of change are geologically quite large; Figure 2.2 shows that plates move at time-averaged speeds of up to several cm per year.

(b) Bathymetry changes because of (i) contraction as rock masses cool (see Section 2.3.2); (ii) deposition of continent-derived sediments; (iii) erosion and submarine canyon formation caused by currents; (iv) isostatic adjustments; and (v) volcanic activity.

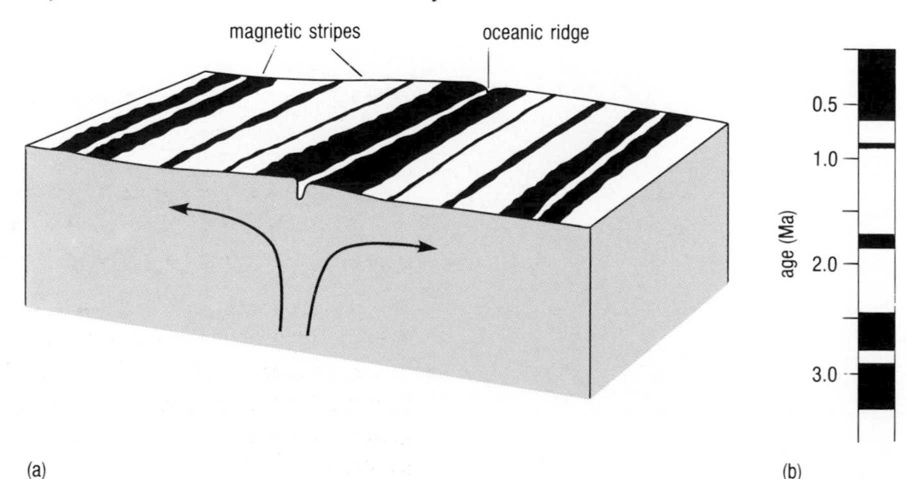

Figure 2.3 (a) Asthenospheric mantle material rises under an oceanic ridge to produce new oceanic lithosphere. As this moves sideways (away from the ridge axis), it cools past the Curie point of magnetic minerals in the rocks and so 'freezes in' the polarity of the Earth's field that prevails at the time.
(b) A simplified polarity time-scale for the past few million years, showing alternations of normal (black) and reversed (white) polarity of the Earth's magnetic field.

2.1 THE MAIN FEATURES OF OCEAN BASINS

Figure 2.4 shows that nearly half of the Earth's solid surface lies within two quite well-defined limits of altitude (0–1 km) and depth (4–5 km), and that a significant proportion lies within a few hundred metres of sea-level.

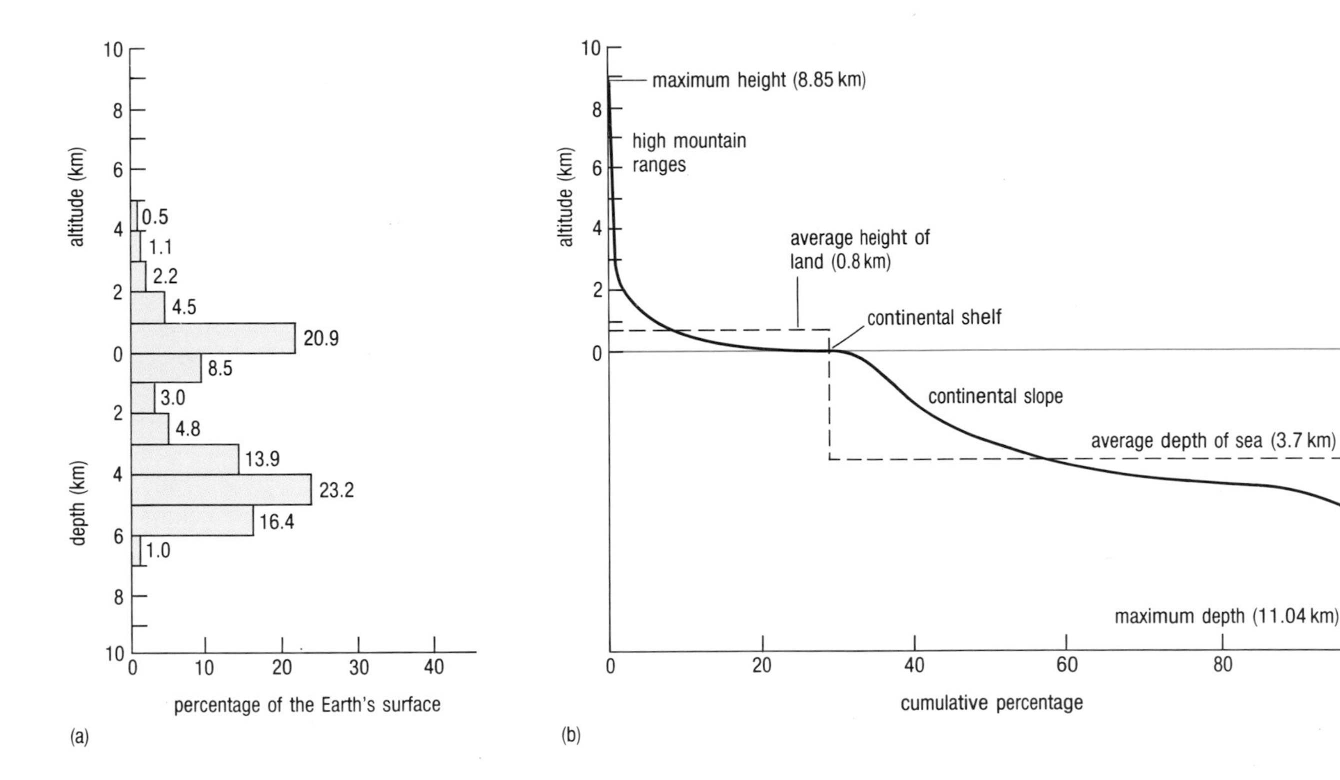

Figure 2.4 The distribution of levels on the Earth's surface.
(a) A histogram showing the actual frequency distribution.
(b) The hypsographic curve: a cumulative frequency curve based on (a). This does *NOT* show the shape of the Earth's surface; it is a curve showing the percentages of the Earth's surface that lie above, below, or between any given levels.

QUESTION 2.1

(a) What percentage of the Earth's surface lies below sea-level?

(b) What would be the effect on this percentage of a 100 m rise in sea-level?

(c) Taking the mean radius of the Earth as 6370 km, what percentage of this radius is represented by the total vertical range in relief of the Earth's solid surface?

The two most striking features of the ocean floor are the world-wide ridge–rift systems traversing all the major ocean basins (and in places encroaching on the continents) and the peripheral (especially circum-Pacific) systems of deep trenches. New oceanic lithosphere is continually being formed by sea-floor spreading at the ridges, hence the alternative terms of spreading axis or constructive plate margin. The lithosphere eventually descends back into the asthenosphere at the trenches which are the surface expression of the inclined planes known as **subduction zones**.

Between the ridges and trenches lie sediment-covered abyssal plains, as well as subsidiary ridges, and a variety of hills and plateaux, some of which break the surface as islands. Bordering the continents are the continental shelves, formed of thick accumulations of sediment. Here, water depths are generally 200 m or less, and the width of the shelf varies greatly from place to place.

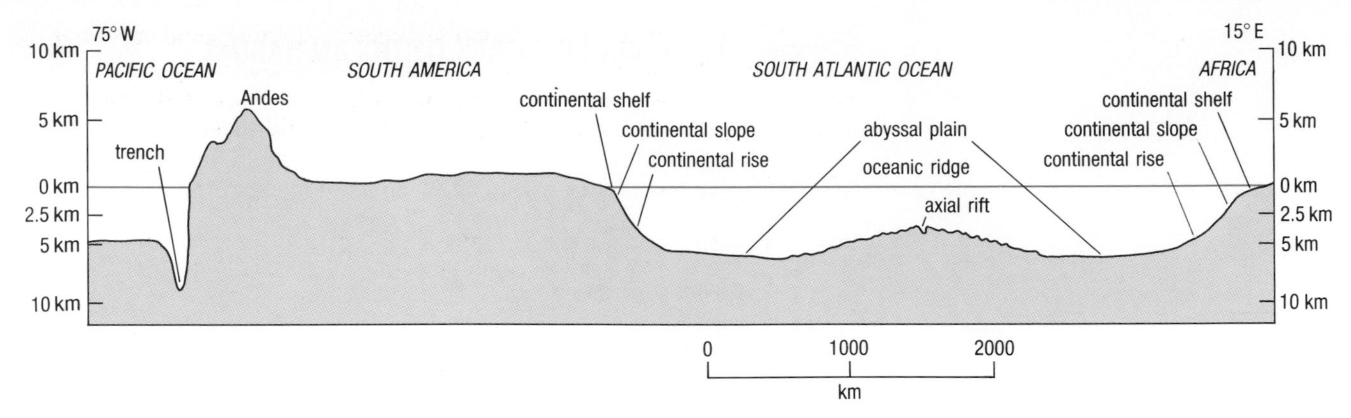

Figure 2.5 Topographic profile to show the surface of the Earth between South America and Africa. Vertical exaggeration × 100.

The relative importance of the main features of the oceans is summarized in Table 2.1. Figure 2.5 is a semi-diagrammatic cross-section (with greatly exaggerated vertical scale) across the South Atlantic and South America (cf. Figure 1.11), to illustrate the main physiographic features listed in Table 2.1. In the following Sections, we discuss the various components of the ocean basins in more detail, beginning at the edges, with continental margins.

Table 2.1 Main features of the principal ocean basins.

	Ocean			
	Pacific	Atlantic	Indian	World ocean
ocean area (10^6 km^2)	180	107	74	361
land area drained (10^6 km^2)	19	69	13	101
ocean area/drainage area	9.5	1.5	5.7	3.6
average depth (m)	3940	3310	3840	3730
area as % of total:				
shelf and slope*	13.1	19.4	9.1	15.3
continental rise*	2.7	8.5	5.7	5.3
deep ocean floor	42.9	38.1	49.3	41.9
volcanoes and volcanic ridges†	2.5	2.1	5.4	3.1
ridges§	35.9	31.2	30.2	32.7
trenches	2.9	0.7	0.3	1.7

* See next Section.

† Volcanic ridges are those related to volcanic island chains that are not part of constructive plate margins, e.g. the Walvis Ridge in the South Atlantic. They do not include island arcs.

§ These ridges correspond to constructive plate margins, e.g. the Mid-Atlantic Ridge.

2.2 CONTINENTAL MARGINS

Even before the advent of plate tectonics, two basic types of continental margin had been recognized and were referred to as Atlantic and Pacific types.

Atlantic-type margins usually have a relatively wide continental shelf and an extensive continental rise (Figure 2.5 and Table 2.1). They have relatively infrequent earthquake activity, and so are termed **aseismic** or **passive margins**. We now know that these develop when a continent is rifted apart and a new ocean forms in between. The continent and adjacent ocean floor are part of the *same* plate (cf. Figure 2.2). Areas of continental crust may become completely isolated by rifting, and form **microcontinents**. Most examples are in the form of submarine plateaux of continental crust that has become stretched and thinned (e.g. the Rockall Bank and Seychelles Plateau). Other micro-continents are large islands such as Madagascar, separated from the main continent by a belt of stretched and thinned crust.

Pacific-type margins typically have a trench at the foot of the continental slope, instead of a continental rise (Figure 2.5). Pacific-type margins occur where oceanic crust is being consumed beneath continental crust at a subduction zone. They are seismically very active (i.e. earthquakes occur frequently) and so are now more commonly known as **seismic** or **active margins**. Here the continent and ocean floor belong to *different* plates (cf. Figure 2.2). Seismic margins can also form at a plate boundary where oceanic crust of one plate is being subducted below oceanic crust belonging to the adjacent plate, in which case they are marked by island arcs (Figure 2.1) – a special case that we shall consider in Section 2.2.2.

2.2.1 ASEISMIC CONTINENTAL MARGINS

Aseismic margins develop by stretching of continental crust and the eventual rifting apart of a continent (to form a new ocean), and are modified by subsequent sediment deposition. A cross-section through such a margin would look something like Figure 2.6. In plan, the shape of the aseismic margin is related to the trend of the original rift. In cross-section, the shape is determined partly by the amount of stretching, faulting (rifting) and associated crustal subsidence, partly by the amount of volcanism associated with the rifting, and partly by the extent of sediment deposition to form the shelf–slope–rise zone. All of these vary considerably from place to place, so that although the basic features can generally be discerned, aseismic margins may differ enormously in detail.

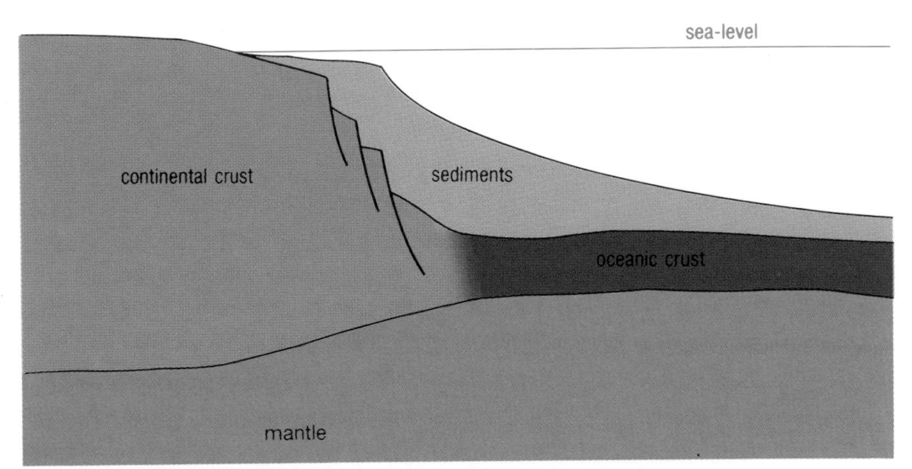

Figure 2.6 Highly diagrammatic cross-section with great vertical exaggeration and not to scale, through an aseismic continental margin. Details of the surface profile are found in Figure 2.7 and related text. Note that this cross-section does not extend deep enough to reach the base of the plate. All the mantle shown belongs to the lithosphere.

QUESTION 2.2

(a) Continental crust is typically thinner than average at aseismic continental margins. Why should thinned (stretched) continental crust subside?

(b) Looking at Figures 2.6 and 2.7, approximately where would you place the transition between continental and oceanic crust?

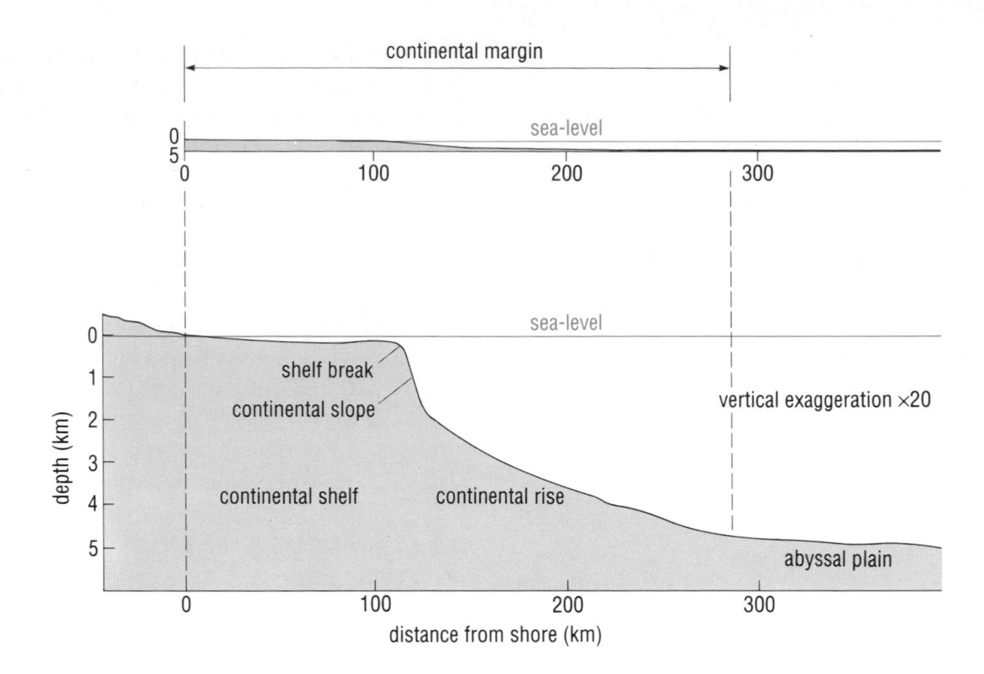

Figure 2.7 One possible configuration of an aseismic (or passive) continental margin, showing the continental rise in relation to shelf and slope. The transition from continental slope to rise is at about 2 km depth, on average.

Because the continental margin is thinned and stretched prior to and during rifting, its surface is below sea-level (as a result of isostasy). Aseismic continental margins are therefore flooded by **shelf seas**, which are up to hundreds of kilometres across. Some shelf seas, such as the North Sea or Hudson Bay, are partly surrounded by land and have long had their own names. Other shelf seas are bounded on one side by the continental mainland and on the other by the open ocean; these correspond to the areas shown in grey in Figure 1.11, many of which lack well-known names. Good examples occur along the east coast of North America and to the north of Australia.

Figure 2.6 shows that the sediments forming most of the **continental shelf**, and therefore forming the floor of the overlying shelf sea, must rest mainly on thinned continental crust (cf. Question 2.2(b)). The width of the continental shelf can be as much as 1500 km, and the surface is generally flat, with an average gradient of only 0.1° (Figure 2.7). However, in such places there may be transient undulations in the form of sand-waves and banks up to a few metres high, generated by the action of currents.

Water depth at the edge of the continental shelf, and therefore at the outer limit of the shelf sea, varies from 20 to 500 m, but averages around 130 m. Here there is a steepening of gradient called the **shelf break** (Figure 2.7), beyond which is the **continental slope**. Continental slopes are usually between 20 and 100 km wide, and between 1.5 and 3.5 km deep at their base. The slope has a much higher gradient than the shelf, averaging about 4° – very steep for submarine relief where gradients tend to be low compared with those on land. Aseismic margins formed by recent rifting of a continent have even steeper continental slopes. In the Gulf of California, for example, which began opening only about 4 Ma ago, the slopes are steeper than 20° because they are yet to be modified by the erosion and sedimentation processes that have been active along the continental margins of the Atlantic for over 100 Ma.

In many areas, such as the Western Approaches (south and west of the British Isles) and the Atlantic coast of North America, the continental slope is traversed by **submarine canyons**. These act as channels for the transport of sediment to the deep ocean. Most submarine canyons start on the continental shelf, commonly (though not invariably) near the mouths of major rivers. They are generally V-shaped in cross-section and resemble river valleys on land. They have been eroded by **turbidity currents** consisting of sediment and water mixtures that are denser than seawater and flow fast enough to scour out the canyons, even down gentle slopes. Turbidity currents are generated by earthquakes and storms over the continental shelf, which trigger the collapse of unstable accumulations of sediment near the shelf edge.

Once the turbidity currents reach the foot of the continental slope, they slow down and begin to deposit their load of suspended sediment, building up the sediment fans that form the **continental rise**. The rise has a much gentler gradient than the slope, about 1° on average. It may be up to 600 km wide, depending partly on the strength and frequency of turbidity currents and partly on the erosive power of currents in the deep ocean circulation. The surface of the continental rise may also be cut by channels along which turbidity currents flow across the rise to the abyssal plains.

2.2.2 SEISMIC CONTINENTAL MARGINS AND ISLAND ARCS

Seismic continental margins are usually associated with an oceanic trench, marking the site of subduction of the oceanic lithosphere into the asthenosphere. These are destructive plate margins.

Trenches may occur next to:

1 continental margins with volcanic coastal mountain ranges, where the oceanic lithosphere on one plate is subducted beneath continental crust on another plate;

2 island arcs, where oceanic lithosphere is subducted beneath oceanic lithosphere on an adjoining plate.

From what you have read so far, what do you think is likely to happen to the sediments carried by turbidity currents flowing down seismic margins, and how do you think the continental slopes of seismic margins differ from those of aseismic margins?

Sediments carried by turbidity currents are generally trapped by trenches at the foot of the continental slopes. As a result, continental slopes at seismic continental margins tend to be steeper than those at aseismic margins (cf. pp. 30–31).

These contrasts are well displayed in the 7000 km-long Peru–Chile trench system, which marks the subduction zone for two plates, the wholly oceanic Nazca Plate and an oceanic part of the Antarctic Plate, beneath the western coast of South America (Figures 2.2 and 2.5). A high range of partly volcanic mountains – the Andes – runs parallel to the coast. Associated with the subduction zone are intense seismicity and volcanism, and a characteristic gravity anomaly pattern (negative at the trench, but positive over the volcanic mountain arc) that characterizes destructive plate margins.

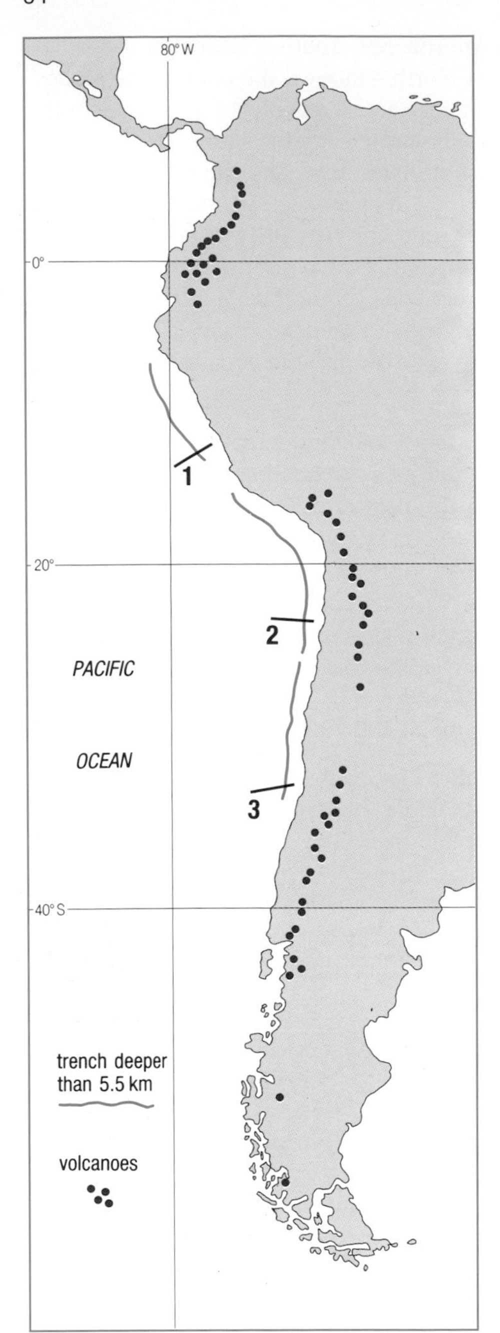

Figure 2.8 The western margin of South America, showing segments of the Peru–Chile Trench that are deeper than 5.5 km. The red dots are active volcanoes. Lines 1–3 correspond to the bathymetric profiles in Figure 2.9.

QUESTION 2.3 Examine Figures 2.8 and 2.9.

(a) How do (i) the width of the continental shelf and (ii) the shape of the continental slope compare with those typical of aseismic margins?

(b) Is the gradient of the continental slope in general consistently greater than anything you might find on aseismic margins?

In some cases (e.g. profiles 1 and 2 in Figure 2.9), differences in the depth and width of the trench are possibly linked with the rate of subduction. For example, fast subduction could result in a deeper and narrower trench. However, there can be other causes of differences in trench profile.

To what would you attribute the differences in the actual trench region illustrated in profiles 2 and 3 of Figure 2.9?

The flat floor of the trench shown in profile 3 must be a result of sediment filling the trench. So why do all the profiles not show similar flat floors built up by thick sediments, which the high relief and close proximity of the Andes would be expected to provide?

At least so far as profiles 2 and 3 are concerned, one possible explanation lies in present climatic conditions:

The Atacama desert in Northern Chile has less than 0.01m of average annual rainfall, so there is virtually no transport of sediment to the ocean by rivers – and the trench in profile 2 reaches a depth of more than 8 km. Further south, however, annual rainfall rises to over 4 m, enabling large amounts of sediment to be transported to the trench, completely filling it south of about 50° S.

Sediments within trenches generally show evidence of deformation, caused by tectonic movements. Extensional (or 'normal') faulting is characteristic of the outer, oceanic, face of the trench wall, and results from warping of the oceanic lithosphere as it bends downwards into the subduction zone. Inner trench walls often carry slices of tightly folded and thrust-faulted (reverse-faulted) sediments, which have been scraped off the descending oceanic plate so that the inner wall builds out progressively oceanwards, forming what is known as an **accretionary prism**. These features are illustrated in Figure 2.10.

Only some of the sediment in the trench is transferred onto the inner trench wall in this way. The remainder is carried down into the mantle by the descending plate. The relative proportions of material scraped off and subducted vary considerably from place to place. Dehydration of subducted oceanic lithosphere and its sediment load leads to partial melting in the overlying mantle (Figure 2.1) and to the building up of a volcanic mountain chain (like the Andes) on the continental crust of the over-riding plate, parallel to the plate margin, and usually no more than a couple of hundred kilometres from it.

QUESTION 2.4 If you look at Figure 2.10, you will see that the ocean-floor sediments forming the accretionary prism have been scraped off as slices or wedges. Would you expect the oldest-formed wedges to be at the top or bottom of the sequence forming the inner trench wall?

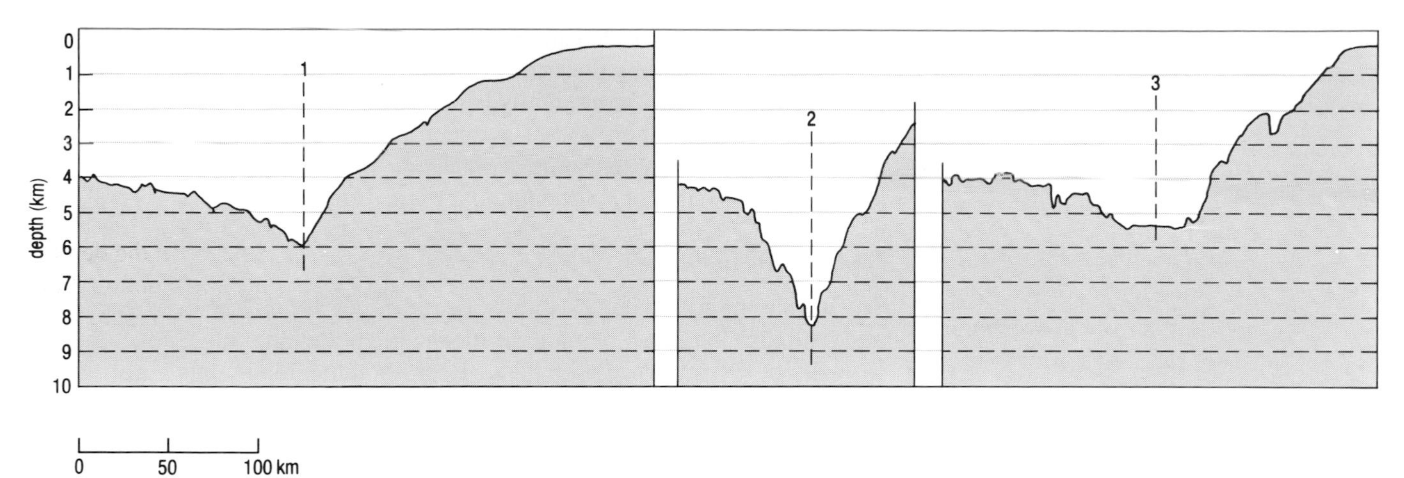

Figure 2.9 Bathymetric profiles across the Peru–Chile Trench at locations given in Figure 2.8. The vertical broken lines indicate the position of the deepest part of the trench corresponding to the blue line in Figure 2.8 (opposite). The vertical exaggeration is ×25.

Island arcs are similar to Andean-type mountain chains, but occur when oceanic lithosphere of one plate is subducted beneath oceanic lithosphere (rather than continental lithosphere), of another. A trench occurs where the two plates meet, and an island arc builds up volcanically on the over-riding plate, above the subduction zone (Figure 2.1). Although island arcs initially form on ordinary oceanic crust, by the time they have reached maturity the extent of magmatic intrusions into the crust and volcanic deposits on top of the crust is such that the crust underlying the arc has become more continental than oceanic in character.

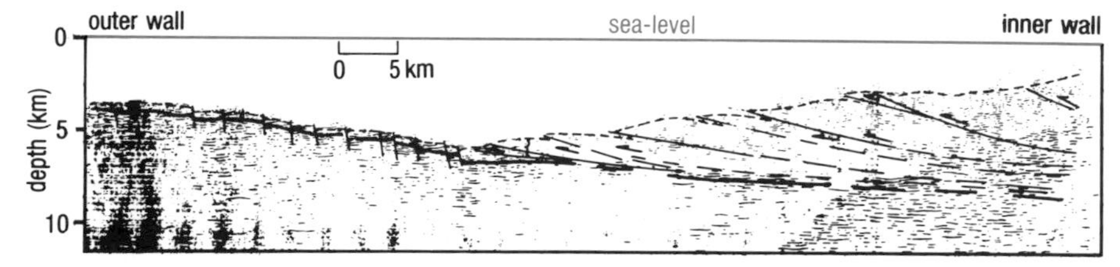

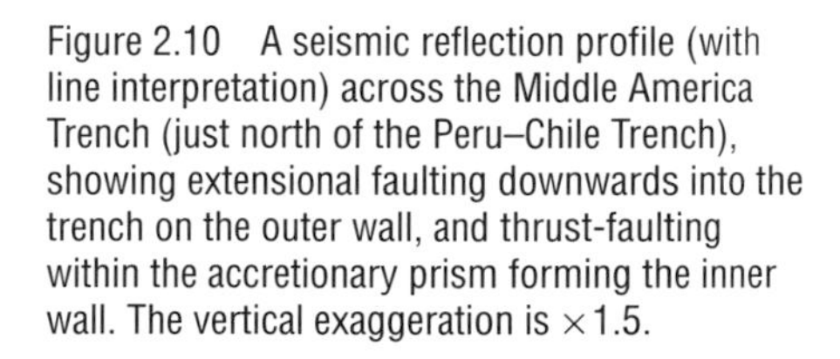

Figure 2.10 A seismic reflection profile (with line interpretation) across the Middle America Trench (just north of the Peru–Chile Trench), showing extensional faulting downwards into the trench on the outer wall, and thrust-faulting within the accretionary prism forming the inner wall. The vertical exaggeration is ×1.5.

Interestingly, a small ocean basin usually opens on the side of the island arc away from the trench. This is called a **marginal basin** or **back-arc basin**. The occurrence of sea-floor spreading in what might simplistically be expected to be a compressional environment (because it is close to where two plates are converging) indicates that plate motions are associated with a complex interplay of forces. In parts of the western Pacific, several sets of island arcs and marginal basins have formed (cf. Figures 1.11 and 2.1), by progressive splitting apart of older arcs.

We now leave the continental margins and subduction zones and move into the deep oceans, to examine the other end of the sea-floor spreading conveyor belt, where oceanic lithosphere is generated.

2.3 OCEAN RIDGES

These are the most important physiographic features of the ocean basins. Most are sites of formation of new oceanic lithosphere, and so are described as constructive plate margins or spreading axes. Elsewhere, you may find the term 'spreading centre' used in place of spreading axis. They are also sometimes referred to as mid-ocean ridges, though this is not an apt description for those in the Pacific (Figure 2.2).

Basaltic magma is extruded and intruded along the median zone of ridges, a process accompanied by the separation of the flanking lithospheric plates, so that new ocean floor is continually being formed. The rate of sea-floor spreading is usually more or less symmetrical about the ridge, adding to the plates on either side at the same average rate, and the direction of spreading is usually approximately perpendicular to the ridge.

2.3.1 RIDGE TOPOGRAPHY

The physiographic characteristics of ocean ridges are related to the spreading rate (and also to the temperature of the mantle below the ridge, a factor that we will not be concerned with here). Representative topographic profiles across a slow-spreading ridge (the Mid-Atlantic Ridge, 1–2 cm yr^{-1*}) and a fast-spreading ridge (the East Pacific Rise, 6–8 cm yr^{-1*}) are shown in Figure 2.11.

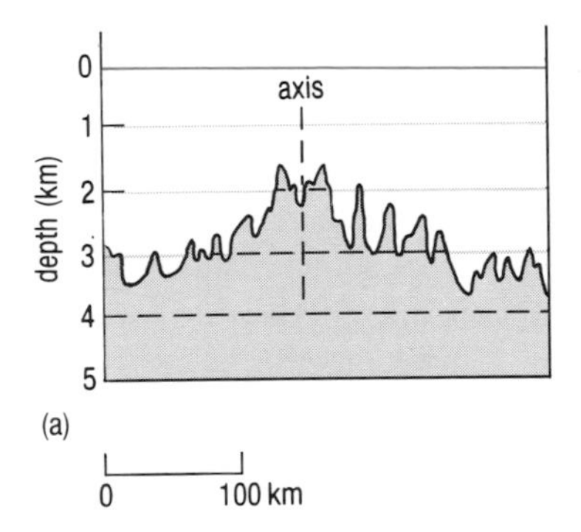

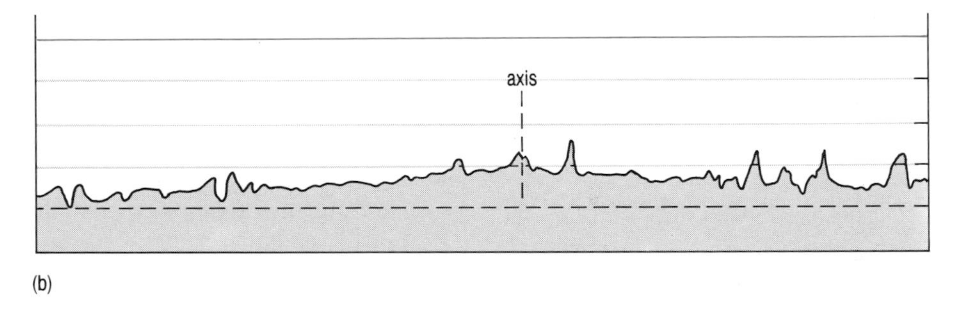

Figure 2.11 Representative east–west topographic (bathymetric) profiles across the Mid-Atlantic Ridge and across the East Pacific Rise (see Question 2.5). The vertical exaggeration is ×50.

QUESTION 2.5 The main features of these two ridges are summarized below. Read this summary, refer to the profiles in Figure 2.11, and then answer (a)–(d).

The Mid-Atlantic Ridge has a median valley about 25–30 km wide and 1–2 km deep, but on the East Pacific Rise a median valley is either absent or narrow (usually < 1 km wide) and shallow (usually <100 m deep). Average gradients away from the ridge crest are of the order of 1 in 100 for the Mid-Atlantic Ridge, and closer to 1 in 500 for the East Pacific Rise.

(a) Which profile is of the Mid-Atlantic Ridge and which is of the East Pacific Rise?

(b) Which ridge has the rougher surface relief?

(c) How do the average gradients of ocean-ridge flanks (given above) compare with those of continental margins?

(d) The Carlsberg Ridge in the northern Indian Ocean has an average spreading rate of 1–2 cm yr^{-1}. Which profile in Figure 2.11 should it more closely resemble?

Profiles across the median valleys of three ocean ridges (Figure 2.12) suggest that these are rift valleys formed by extensional faulting on fault planes inclined towards the axis. On the flanks beyond the valleys there are some outward displacements, along faults inclined away from the axis (e.g. the right-hand end of profiles 2 and 3). This block faulting is responsible for much of the rough topography of ridge crests.

* Unless stated otherwise, spreading rates are half-rates, i.e. the rate at which a plate moves away from the spreading axis rather than away from the other plate.

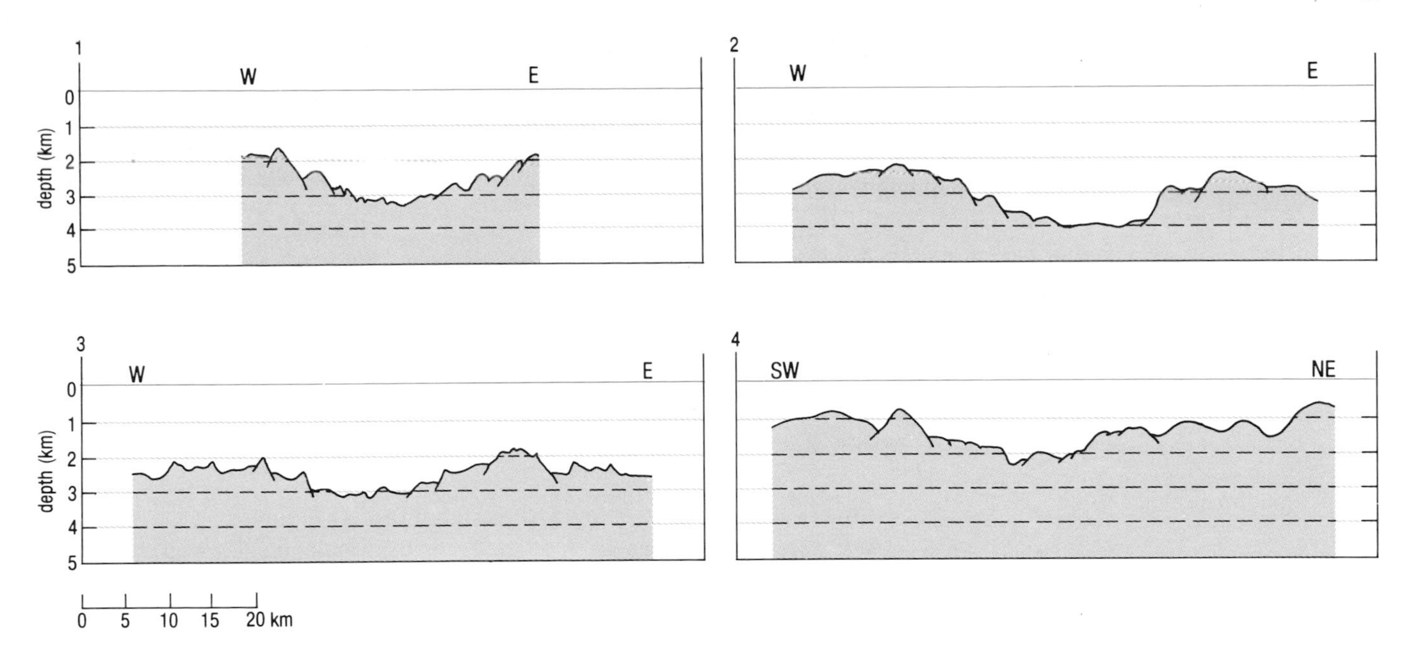

Figure 2.12 Detailed topographic profiles across the median valleys of the Mid-Atlantic Ridge at 47° and 22° N (Profiles 1 and 2); the Gorda Ridge, north-east Pacific Ocean (3), and the Carlsberg Ridge, north-west Indian Ocean (4). The vertical exaggeration is ×5 throughout.

2.3.2 AGE–DEPTH RELATIONSHIPS ACROSS RIDGES

One of the most interesting features of ocean ridges is that the depth of water increases systematically with the age of the lithosphere, and hence with increasing distance from the ridge (Figure 2.13). The reason for this relationship is that the lithosphere is hottest near the ridge axis, where it is forming, and is therefore least dense and most buoyant. As the lithosphere cools over millions of years while moving away from the ridge, it contracts, becomes less buoyant and subsides. The **age–depth relationship** embodied in the curves in Figure 2.13 can be used to determine the approximate age of oceanic crust of known depth, and can help in the construction of palaeobathymetric (ancient bathymetry) maps which show how ocean basins developed.

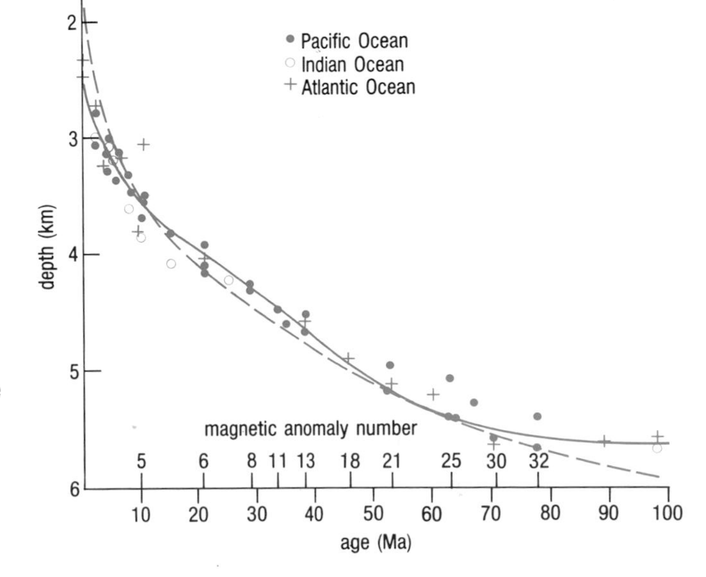

Figure 2.13 Observed and theoretical relationships between the depth to the top of the oceanic crust and its age. The solid line is a best-fit curve through observed points. The dashed line is a theoretical elevation curve, calculated on the assumption that an increase of depth with age is caused by the thermal contraction of the lithosphere as the plate cools on moving away from the ridge axis. Magnetic anomaly numbers refer to the linear magnetic stripes on the ocean floor, which are arranged symmetrically about ridge axes.

There are some limitations, however:

1 The age–depth relationship cannot be used for oceanic lithosphere older than about 100 Ma, as by then the lithosphere has lost most of its heat of formation and approached thermal equilibrium, i.e. the curve has levelled off.

2 The depth of the ridge itself varies with the rate of spreading (fast-spreading axes usually being deeper than slow ones); the average depth to the axis of the East Pacific Rise is about 2.7 km, that to the Mid-Atlantic Ridge axis is about 2.5 km (in this respect, the bathymetric profiles in Figure 2.11 are not representative).

2.4 TRANSFORM FAULTS AND FRACTURE ZONES

Ridge axes may be offset along **transform faults** (Figure 2.1), where oceanic plates slip sideways past each other. Transform faults lie along arcs of circles about the pole of relative rotation between the plates on either side, and they are parallel to the direction of relative movement of the plates (Figure 2.14). Transform faults are so called because they link two points at the ends of successive ridge segments where spreading motion is transformed into sideways slip.

Because of the relative motions of the plates on either side of them, earthquakes occur along transform faults. It is important to realize that a transform fault linking the ends of two offset ridge segments ends where it reaches the ridge axis. Beyond those points, the topographic extension of a transform fault lies within a single oceanic plate, has no relative lateral

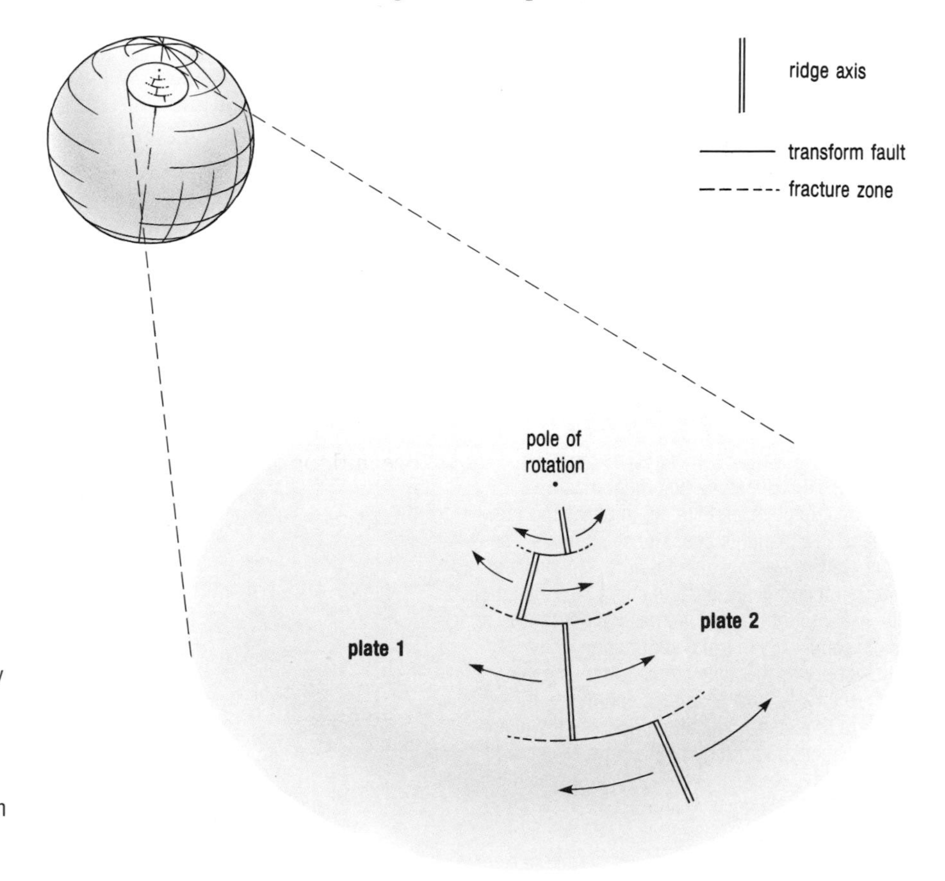

Figure 2.14 Both transform faults (heavy lines) and their inactive extensions, fracture zones (dashed lines), are 'small circles' centred on the pole of relative rotation of the two plates. The spreading rate is related to the angular rate of plate separation and distance from the rotation pole. Spreading rates therefore increase gradually with distance from the rotation pole, as indicated by different lengths of arrows. Parts of just two plates are shown here. Motion of *any* two plates can be described with reference to a particular pole of rotation. *Note:* 'small circles' are circles on the face of the globe *not* including the Equator, lines of longitude or any other 'great circle'.

motion across it and shows little or no major seismic activity. Such a feature is called a **fracture zone** (Figure 2.14).

If the original rift between the two continents is at an oblique angle to the spreading direction, the spreading axis will usually develop as a series of offset ridge segments, as illustrated in Figure 2.15. Weak zones in rifting continents probably control the sites where the more important transform faults develop, as may have happened when South America and Africa split apart to form the equatorial Atlantic Ocean.

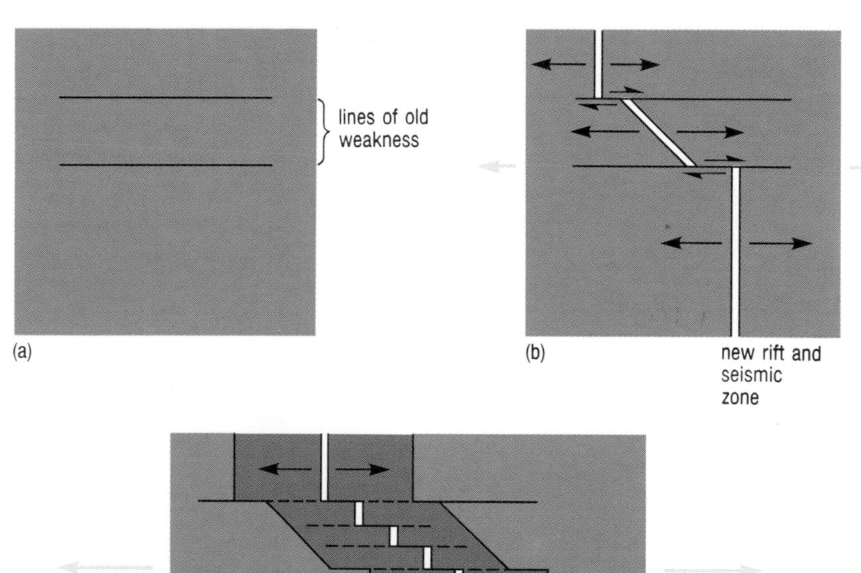

Figure 2.15 The break-up of a continent, with ancient lines of weakness controlling the development of major transform faults, which offset the ridge segments and become (mainly aseismic) fracture zones outside the zone of active spreading. Ridges are normally perpendicular to spreading directions, though not invariably so. (*Note:* Key as for Figure 2.14.)

Transform faults and fracture zones are usually expressed as bathymetric features such as scarps and clefts on the ocean floor. The age–depth relationship described in Section 2.3.2 explains the origin of these scarps (Figure 2.16): on opposite sides of a transform fault or fracture zone, the lithosphere is of different ages and therefore lies at different depths. Fracture zones are not totally aseismic, because adjustments in the relative depth of ocean floor on either side of a fracture zone must occur as the lithosphere cools, contracts and subsides at different rates on either side. Generally, this produces only small earthquakes. Near continental margins, there is an additional cause of vertical displacements, accompanied by minor earthquakes: sediment brought in by rivers and turbidity currents and accumulating at different rates on either side of a fracture zone can lead to differential loading, and greater subsidence on one side of the fracture than the other.

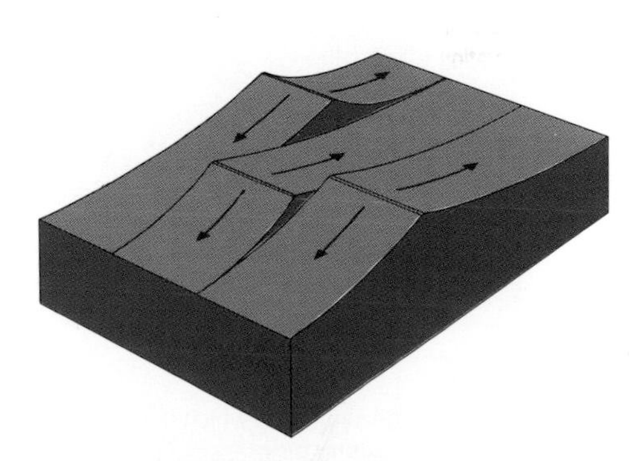

Figure 2.16 The elevation of the sea-floor is different on opposite sides of transform faults and fracture zones, so that escarpments are formed.

Transform faults differ fundamentally from the transcurrent faults marking lines of lateral movement that are familiar to geologists on land: the actual movement along transform faults is in the *opposite* direction from that suggested by the ridge offsets, because of spreading at the ridge axis (Figures 2.14–2.16).

Large transform faults are also known as **conservative plate margins** (e.g. the San Andreas Fault in western North America and the New Zealand Alpine Fault, Figure 2.2). This is because there is normally no component of subduction or spreading along them, so oceanic crust is neither created nor destroyed. One of the best-known exceptions to this general rule is the plate boundary in the Atlantic Ocean between the European and African Plates (Figure 2.2). This is a major transform fault, but it is not a wholly conservative plate margin. At its western end, near the Mid-Atlantic Ridge, it has a small spreading component as well as the dominant transcurrent (lateral) motion, and there is some leakage of magma from the upper mantle, leading to eruption of basaltic lava along it. This type of fault is known as a 'leaky transform'. Further east, nearer to Gibraltar, it is a typical conservative margin, and the only movement is transcurrent. Where it enters the Mediterranean it acquires a significant component of subduction, again superimposed on the dominant lateral movement. Smaller transform faults can also show these features, especially a small spreading component, and leaky transform faults are not uncommon along ocean ridges (Figure 2.17).

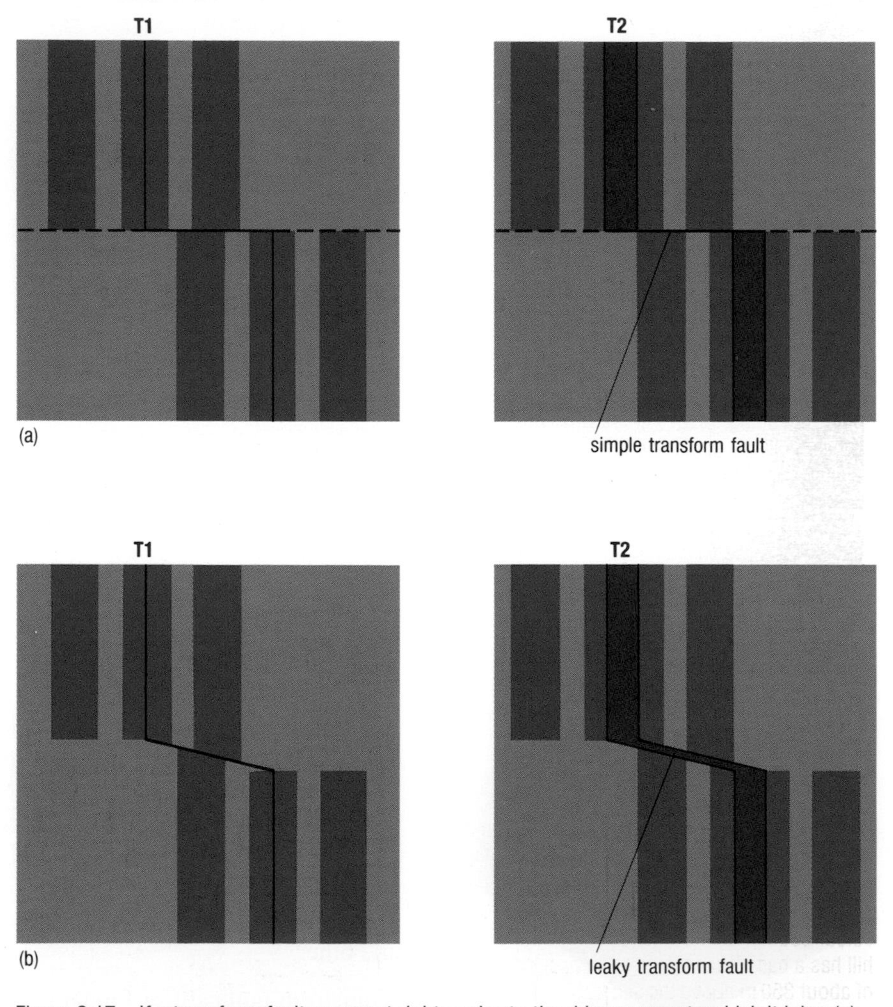

Figure 2.17 If a transform fault occurs at right angles to the ridge segments which it joins (a), then it lies parallel to the direction of sea-floor spreading, and spreading progresses (T1 to T2) without generation of new sea-floor along the transform fault. However, if a transform fault occurs oblique to the direction of spreading (b), then the fault has to be 'leaky', to fill the space left as the plates diverge.

Transform faults and fracture zones are simple in principle, but complex in detail, as revealed by surveys using side-scan sonar techniques and swath bathymetry (Section 1.2) and submersibles, in addition to geophysical measurements. For example, displacement may occur either along a single main fault (the 'principal transfer displacement zone'), or along a complex of branching smaller faults.

Minor offsets (less than about 10 km) of the linearity of a spreading axis can be accommodated without transform faults, and several types of such 'non-transform offset' are now recognized. The character of these differs between fast- and slow-spreading axes, and they are described in Chapter 4.

2.5 THE DEEP OCEAN FLOOR

Between the continental margins and the ocean ridges lies the deep ocean floor, representing about 42% of the total oceanic area (Table 2.1). This is a region of varied topography (bathymetry) and relief (Figure 1.11), comprising featureless abyssal plains, ridges, seamounts, plateaux or banks (which may be microcontinents or products of submarine volcanism), and long narrow clefts (the surface expression of major fracture zones).

2.5.1 ABYSSAL PLAINS

Abyssal plains have average gradients of less than 0.05°: a change in height of less than 1 m per km. This is flatter than any other ocean feature and much flatter than most land areas. However, the flatness is frequently interrupted by abyssal hills and seamounts. Many rise abruptly from the plain, suggesting that their bases are buried by sediment. Strictly speaking, only those exceeding a kilometre in height are classifiable as **seamounts** (Section 2.5.2), whereas smaller features are described as **abyssal hills** (Figure 2.18). Seamounts are generally volcanic in origin, as are many abyssal hills, though abyssal hills can also be formed by faulting. The gradient of the plain decreases away from the continental rise until it reaches the distant foothills of the spreading ridge, known as the abyssal hill province, where the fault-controlled topography of the ridge is only partly buried by sediment.

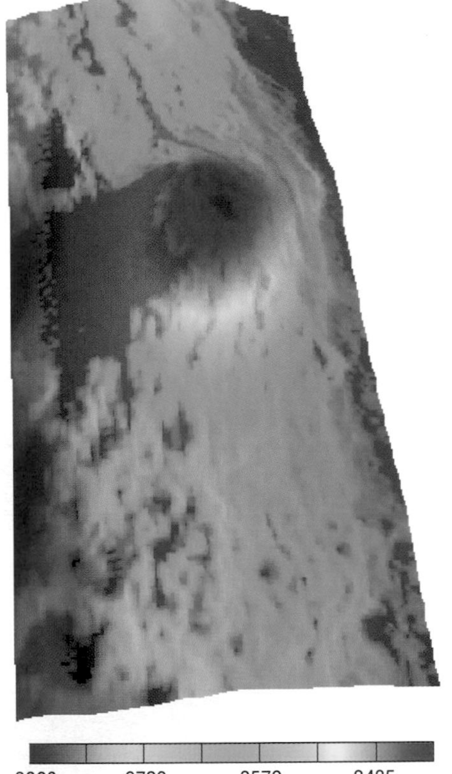

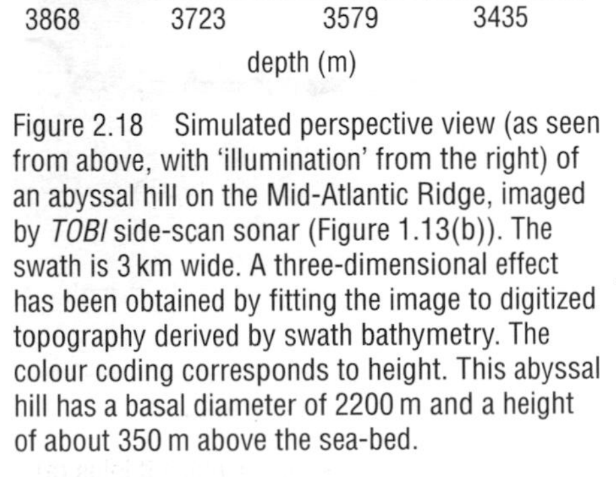

Figure 2.18 Simulated perspective view (as seen from above, with 'illumination' from the right) of an abyssal hill on the Mid-Atlantic Ridge, imaged by *TOBI* side-scan sonar (Figure 1.13(b)). The swath is 3 km wide. A three-dimensional effect has been obtained by fitting the image to digitized topography derived by swath bathymetry. The colour coding corresponds to height. This abyssal hill has a basal diameter of 2200 m and a height of about 350 m above the sea-bed.

QUESTION 2.6 Look at Figure 2.19, which summarizes the description we have just given. From which major ocean basin might this profile have come?

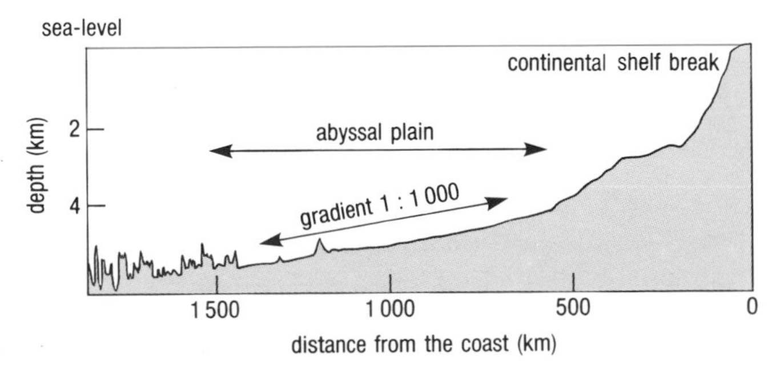

Figure 2.19 A topographic profile across an abyssal plain and continental rise and slope.

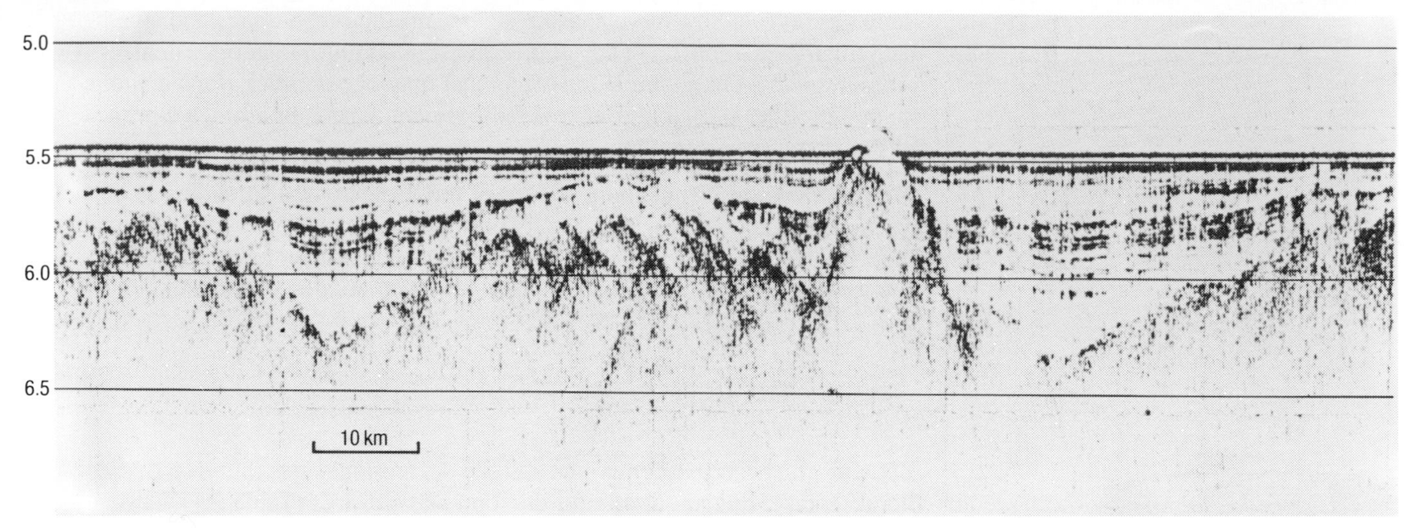

Figure 2.20 A seismic reflection profile across part of the Madeira abyssal plain (north-east Atlantic). The vertical exaggeration is ×20.

Turbidity currents deposit most of their load to build up the sediment wedge of the continental rise, but some of the load is carried further out and, along with pelagic sediments (i.e. those deposited from suspension in the open oceans, see Section 6.1), gradually blankets the rough topography of the oceanic crust that is formed at ridge axes. Figure 2.20 shows the rugged topography that can underlie abyssal plains, and how effective the burial process is, with only the tops of hills remaining exposed.

Would you expect abyssal hills to be as extensively buried by sediment in the Pacific Ocean as in the Atlantic?

The Pacific Ocean is bordered either by trenches or by marginal seas behind island arcs, both of which trap almost all the sediment brought in from continental areas by turbidity currents. In the deep Pacific, sediment cover is thinner than in the Atlantic, being predominantly pelagic rather than continental in origin. Therefore, abyssal hills (like seamounts – see below) tend to be incompletely buried. One estimate is that, of the deep ocean floor in the Pacific, about 90% is hilly terrain and only 10% is smooth abyssal plains.

2.5.2 SEAMOUNTS

The majority of seamounts form as ocean-floor volcanoes, most of which do not extend above sea-level. Those that do are called oceanic islands, but the term seamount is commonly used to describe both islands and subsea volcanoes. Seamounts are found throughout the oceans, but are particularly common in the Pacific, partly because of their less complete burial by abyssal plain sediment.

Figure 2.21 shows the topographic profiles of some seamounts. The angles of the slopes can be as great as 25°, so seamounts are among the steepest features on the ocean floor. Seamounts are roughly circular in plan and some are very large. The largest is the Big Island of Hawaii, whose base on the floor of the Pacific is about 100 km across and lies about 9 km below its highest point (the summit of Mauna Loa volcano).

Many seamounts are of similar age to the surrounding ocean floor, and must have formed on or near a spreading axis. Some once reached above sea-level,

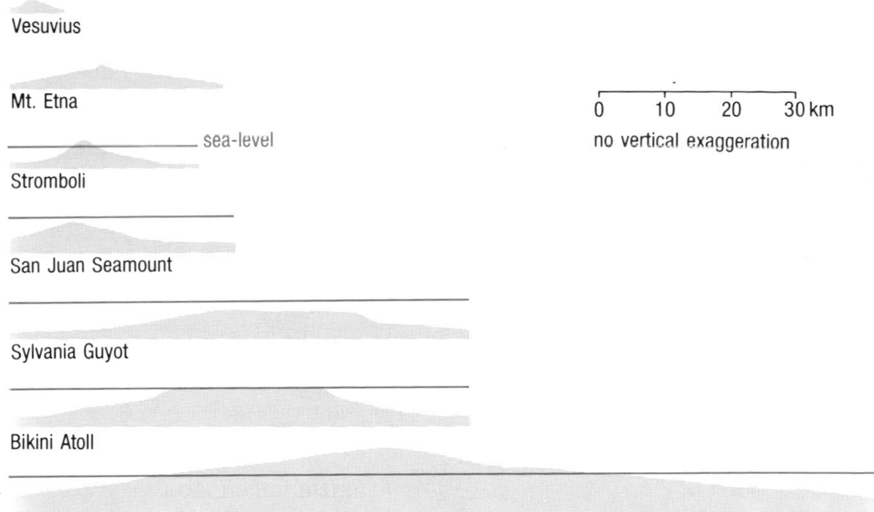

Figure 2.21 Topographic profiles across some on-land volcanoes and some seamounts.

but are now below the surface. This is partly because of the general subsidence of the whole oceanic lithosphere as it moves away from a spreading ridge, and partly because of local isostatic subsidence, in response to the additional load of the seamount itself on the crust. Flat-topped seamounts, or **guyots** (Figure 2.21), are generally believed to have been planed off by wave erosion when the volcanoes were at sea-level (but see also Chapter 4).

The tops of many guyots are well below sea-level, far beyond the reach of the action of waves or surface currents. Would you expect to find a relationship between the depth to the top of guyots and the age of the surrounding sea-floor?

In general, we would expect the summits of older guyots to be further from spreading axes and hence to be at greater depths, having subsided more since their formation (cf. Figure 2.13). However, in tropical regions, many seamounts and guyots have been massively colonized by corals, coralline algae and other wave-resistant encrusting organisms so that reefs and atolls of calcium carbonate (limestone) have been built up as they have subsided, maintaining their summits near sea-level.

Where seamounts have formed by volcanism *away* from the spreading axis, on older oceanic crust which had already cooled down since its formation, the surrounding crust has been reheated, giving rise to a thermal anomaly. Such crust is warmer and thus more buoyant than would be expected for its age, and so plots *above* the age–depth curve of Figure 2.13.

2.5.3 THE DISTRIBUTION OF SUBMARINE VOLCANOES

Many submarine volcanoes appear to be randomly scattered about the ocean floor (Figure 1.11). Others, however, are arranged in linear chains. Figure 2.22 shows four linear chains of volcanic islands and seamounts in the Pacific Ocean, all aligned roughly NW–SE. Age determinations suggest a general progression in the ages of individual volcanoes from older to younger south-eastwards along each chain. This has lead to the formulation of the '**hot spot**' concept, according to which each linear chain is formed as a result of an oceanic plate (in this case, the Pacific Plate) moving over a narrow heat source, or **mantle plume**. This consists of pulses of material rising from deep in the mantle to a point fixed relative to the Earth as a whole, and across which the lithospheric plate moves (Figure 2.22).

There is a regular progression of ages along the Hawaiian–Emperor Chain, and Figure 2.22 shows that the age range along the Hawaiian Chain is 43 Ma over a distance of 3400 km.

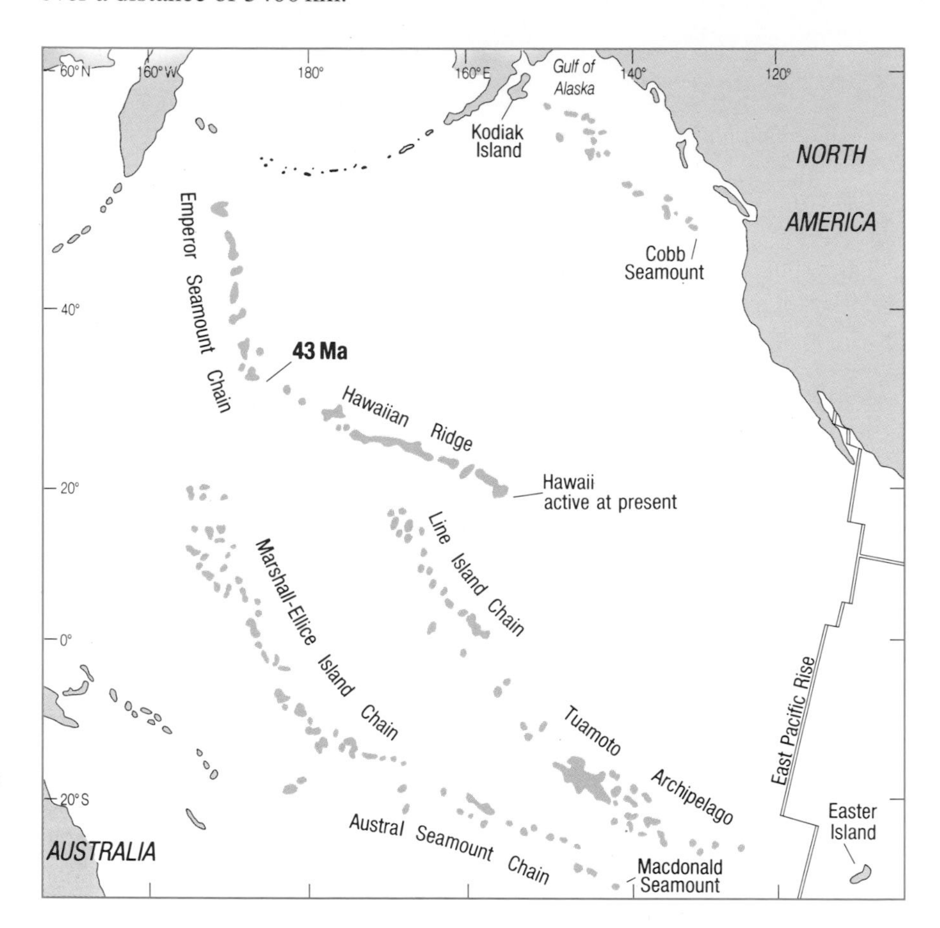

Figure 2.22 Four seamount and island chains in the Pacific Ocean. The youngest volcanoes are at the south-eastern end of each chain, and the age of the seamount at the bend in the Hawaiian–Emperor Chain is shown.

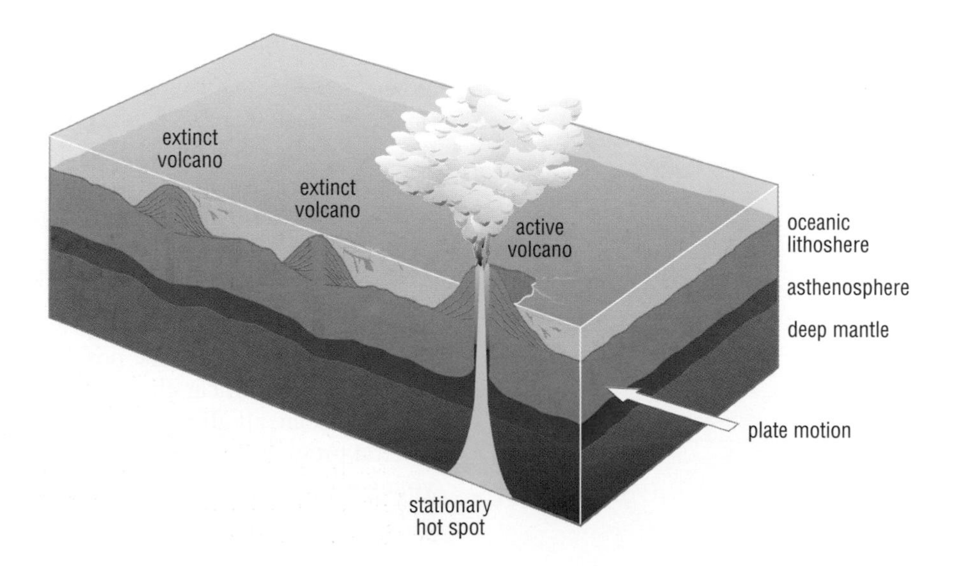

Figure 2.23 Schematic diagram (not to scale) illustrating how a volcanic island chain could be formed by an oceanic plate moving over a stationary hot spot or mantle plume. The age of the islands increases towards the left. New islands will appear on the right as the motion continues.

QUESTION 2.7

(a) What would be the average rate of movement of the Pacific Plate according to the hot-spot model?

(b) Assuming that a constant rate of movement has been maintained, what is the approximate age of the northern end of the 1900 km-long Emperor Chain?

(c) Suggest a reason for the kink in the chain as a whole (at the western end of the Hawaiian sector, the southern end of the Emperor sector).

(d) In fact, the age of the northern end of the Emperor Chain has been dated at about 72 Ma. Assuming that a change in *rate* of plate motion coincided with the change in *direction* of motion you have just postulated, what was the average rate of movement of the Pacific Plate between 72 and 43 Ma ago?

Data from the Hawaiian–Emperor Chain provide strong support for the hot-spot model; but evidence from other chains is more equivocal. For example, there is an overall south-eastward decrease in age within the nearby Line Islands–Tuamoto Chain that is consistent with an origin at a stationary hot spot, but several seamounts in the chain have anomalous ages. In this case, there has evidently been some volcanism not directly related to a Hawaiian-type hot spot.

2.5.4 ASEISMIC RIDGES

These are some prominent and more or less continuous features traversing the deep ocean floor, which in some cases rise to more than 3000 m above it. Examples include the Walvis Ridge and Rio Grande Rise in the South Atlantic, and the remarkably long and straight Ninety-east Ridge in the Indian Ocean (Figure 1.11). The available evidence supports a hot-spot mechanism for the origin of these features.

QUESTION 2.8 What must be the main difference in the character of the volcanism forming aseismic ridges from that forming a seamount/island chain?

2.6 SATELLITE BATHYMETRY – A CASE STUDY

This Section examines how satellite radar altimeter measurements can be used to refine bathymetric maps, notably in parts of the relatively less well-surveyed southern oceans (Figure 1.15). In Section 1.2.1, you saw how the geoid (the surface of equal gravitational potential) varies round the Earth, and thus how, in the absence of winds, tides and currents, the mean sea-level has a surface relief of as much as 200 m. This relief is the result of the effects of both large-scale and small-scale gravitational variations. Thus, there are long-wavelength (> *c*. 500 km) undulations in the geoid which result from deep-seated mass anomalies within the mantle or lower crust, and much shorter wavelength undulations (often referred to as 'geoid anomalies') reflecting shallow structures and sea-floor topography (Figure 1.16). The magnitude of longer-wavelength effects is large enough to obscure the shorter-wavelength features, so the longer-wavelength trends must be filtered out from the altimeter data obtained from the satellite before the information relating to bathymetry becomes apparent. You saw such a global geoid anomaly map derived from *Geosat* and *ERS*-1 data in Figure 1.18. We will begin here by looking at the advances in knowledge gained using the less precise and less detailed *Seasat* data.

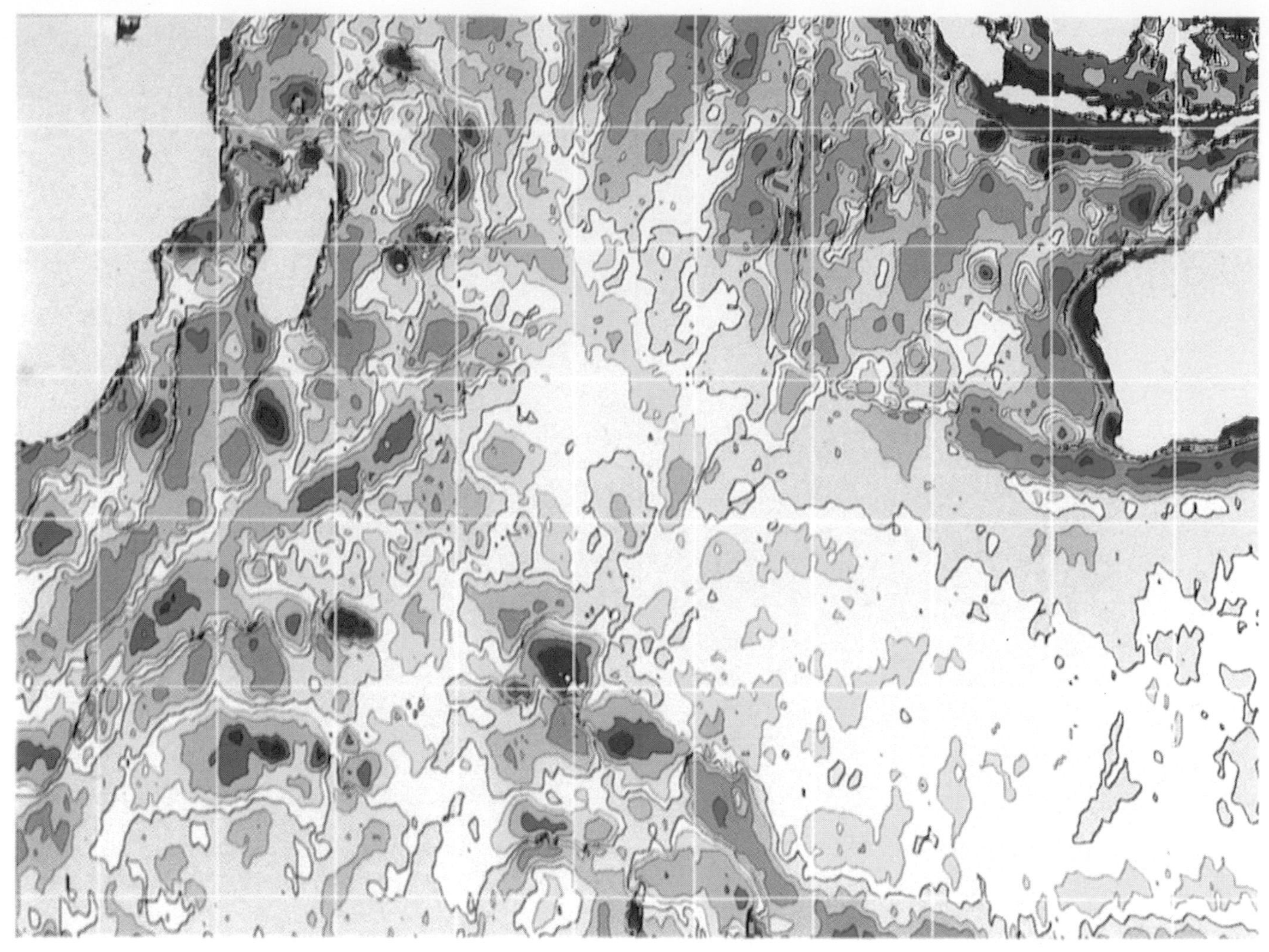

Figure 2.24 (a) Geoid anomaly map of the South Indian Ocean derived from *Seasat* mean surface measurements, processed to remove larger-scale anomalies (more than *c.* 500 km). The contour interval is 1 m, except for the +1 to –1 range, where it is 0.5 m. Darkest blue represents areas lower than –4 m, brightest red represents areas higher than +4 m. The zero contour (heavy black line) is between lightest blue and palest brown. Latitudes and longitudes as in Figure 2.24(b).

Seasat data were collected over 70 days during the 100-day lifetime of the satellite from 5 July to 10 October 1978. Temporal variations related to transient oceanographic effects were averaged out by repeated passes over the same points during different orbits. The *Seasat* altimeter was sufficiently precise to define the distance from the satellite to the sea-surface to within 1 part in 10^7, or about 5 cm, but the height of the satellite in its orbit was known only to within about 1.6 m. However, such imprecisions in satellite orbit appear mostly as long-wavelength variations, and so have little effect on the short-wavelength features relevant to bathymetry. By the use of averaging techniques, *Seasat* data were adequate to produce a map of the marine geoid with a horizontal resolution of about 50 km and an internal precision in height of about 20 cm.

Figure 2.24(a) is a contoured rendering of the geoid anomaly map derived from *Seasat* radar altimetry data covering the southern Indian Ocean, and Figure 2.24(b) is a map of the same area drawn from bathymetric information obtained from previous shipborne observations. Comparison between these shows how satellite information is of great assistance in ocean exploration, but also illustrates some of the pitfalls for the unwary (as you will find in Question 2.9).

Thanks to the altimetry data, hitherto unidentified features can be located, and the bathymetry of known features can be refined. This is especially well exemplified in the vicinity of the Kerguelen Plateau, which was previously

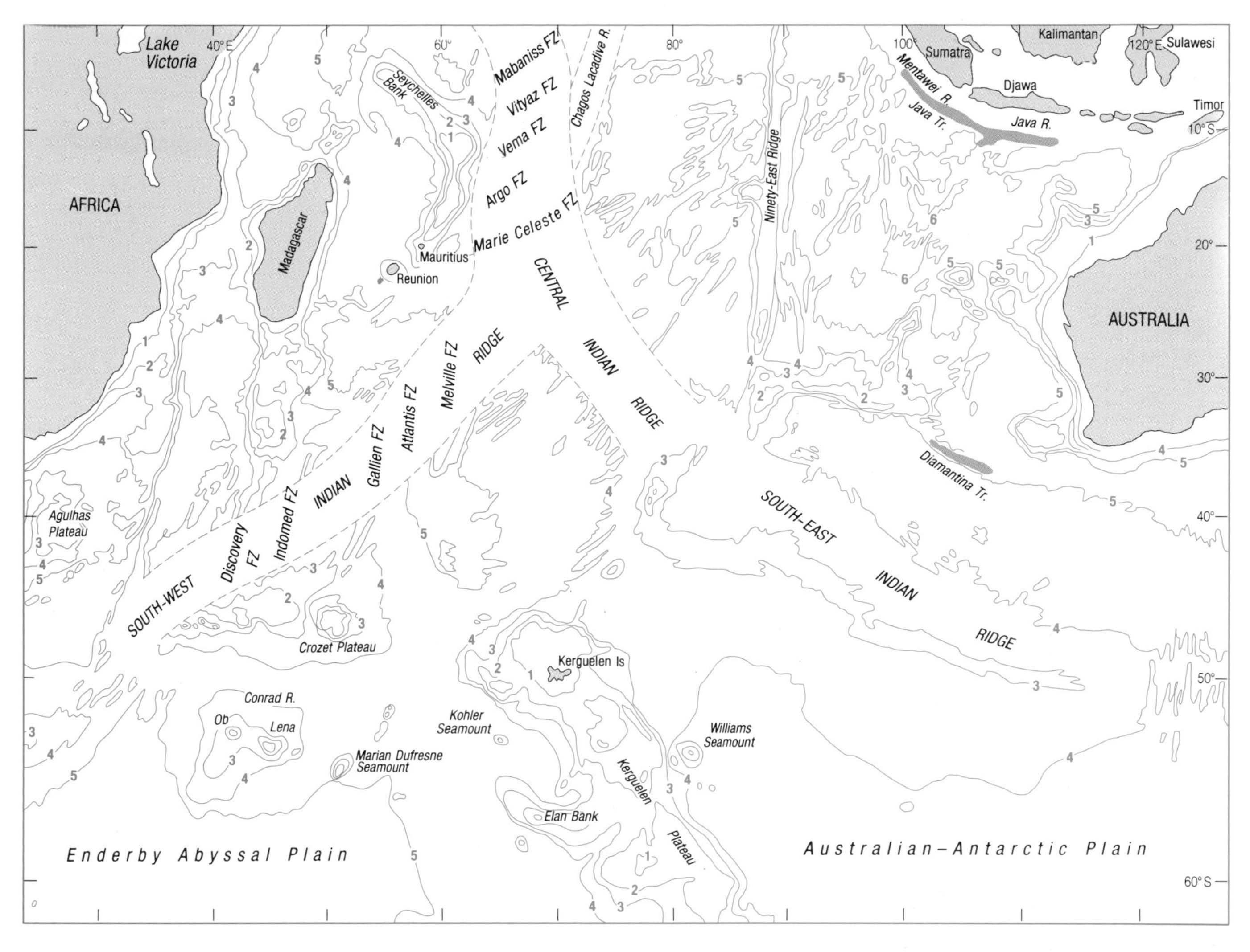

(b) Bathymetric map, compiled from conventional shipborne measurements of the South Indian Ocean. Area north-east of Java Trench not contoured. Note the 'triple junction' where three spreading axes meet at ~25° S, 70° E. Complex topography less than 4 km depth on active ridges (between dashed lines) not contoured. Contour interval = 1 km; blue shaded features are trenches or troughs, and contours are omitted. R = ridge or rise; FZ = fracture zone; Tr = trench or trough.

poorly known, chiefly because ship tracks in this region were too sparse to have defined the features adequately (cf. Figure 1.15). By way of example, two of the newly distinguished features on the geoid anomaly map are:

1 A large trough east of the Kerguelen Plateau, between 54° S and 60° S.

2 An uncharted seamount between Lena Seamount and Marian Dufresne Seamount, at 53° S, 49° E.

QUESTION 2.9

(a) On Figure 2.24(a), what features show up west of the Ninety-east Ridge, and how well do they correlate with the previously known bathymetry?

(b) On Figure 2.24(b), how does the bathymetric expression of the Diamantina Trough relate to its expression on the geoid anomaly map (a)?

(c) How good is the correspondence between the geoid map and the bathymetric chart for the following three plateau regions in the western Indian Ocean: (i) Agulhas Plateau, (ii) Madagascar Plateau (south of the island of Madagascar) and (iii) Mozambique Plateau (between the two, nearer the coast of southern Africa)? What is the size of the geoid anomaly over these features?

(d) In general, how do fracture zones show up on the geoid anomaly maps?

(e) How are the Java Trench and the sea-floor to the south of it expressed on the geoid anomaly map (Figure 2.24(a))?

In Question 2.9(e), you encountered one example of how bathymetry inferred from geoid anomalies alone must be treated with caution. There are two other possible sources of significant error. First, not all bathymetric features have significant geoid anomalies over them. For example, seamounts which formed on young oceanic crust, while its flexural rigidity was still low, will have little effect on the geoid, because the load has been locally compensated for (Figure 2.25(a)). The loading by features formed on older, cooler, and therefore more rigid crust will be spread over wider areas of the plate (Figure 2.25(b)), so the elevation of the geoid locally over the seamount itself will be greater.

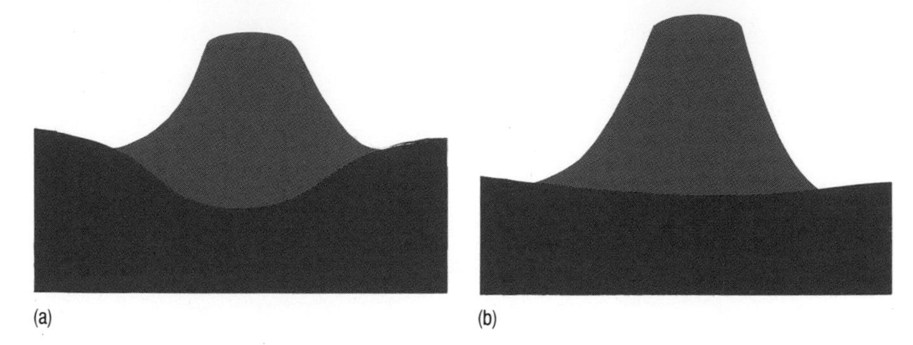

Figure 2.25 (a) A younger and more flexible plate will accommodate the load of a seamount locally. The gravitational anomaly (and hence the geoid anomaly) over it is less than in (b) which shows a seamount formed on an older, more rigid plate.

The second problem is the converse of the first: there may be density anomalies in the crust and upper mantle that have little or no surface expression. Thus, satellite altimetry cannot give unequivocal information about ocean bathymetry. There must always be some control by more conventional shipborne techniques, to confirm or refute the interpretation. We will see an example of this shortly.

In Chapter 1 you saw how the radar altimeter data from the *Geosat* and *ERS*-1 and *ERS*-2 satellites have provided geoid anomaly maps (Figure 1.18) that are even better than those from *Seasat*. This improvement is demonstrated by Figure 2.26 which shows the north-central part of the area shown in Figure 2.24 mapped by both *Geosat* and *ERS*-1 and by *Seasat*. This area includes the 'triple junction' where three spreading axes meet.

QUESTION 2.10

(a) With reference to Figure 2.24, to what do the two geoid anomaly highs in the extreme north-west of Figure 2.26 correspond?

(b) In Figure 2.26, in what way do the expressions of the following differ between the *Seasat* data (bottom) and the *Geosat/ERS*-1 data (top):
(i) transform faults and fracture zones, (ii) median valleys?

Important as the *Seasat* data were in their day, it is clear that the succeeding generation of geoid anomaly data from *Geosat* and *ERS*-1 provide even more information. For example, satellite bathymetry in the form of *Geosat/ERS* data is a powerful tool in mapping sea-floor spreading patterns.

QUESTION 2.11 From your understanding of previous Sections, what is the correlation between spreading rate, topographic relief and geoid anomaly? Bearing this in mind, what do the differences between the appearance of the South-west Indian Ridge and that of the Central Indian Ridge in the *Geosat/ERS*-1 geoid anomaly map in Figure 2.26 suggest about their relative rates of spreading?

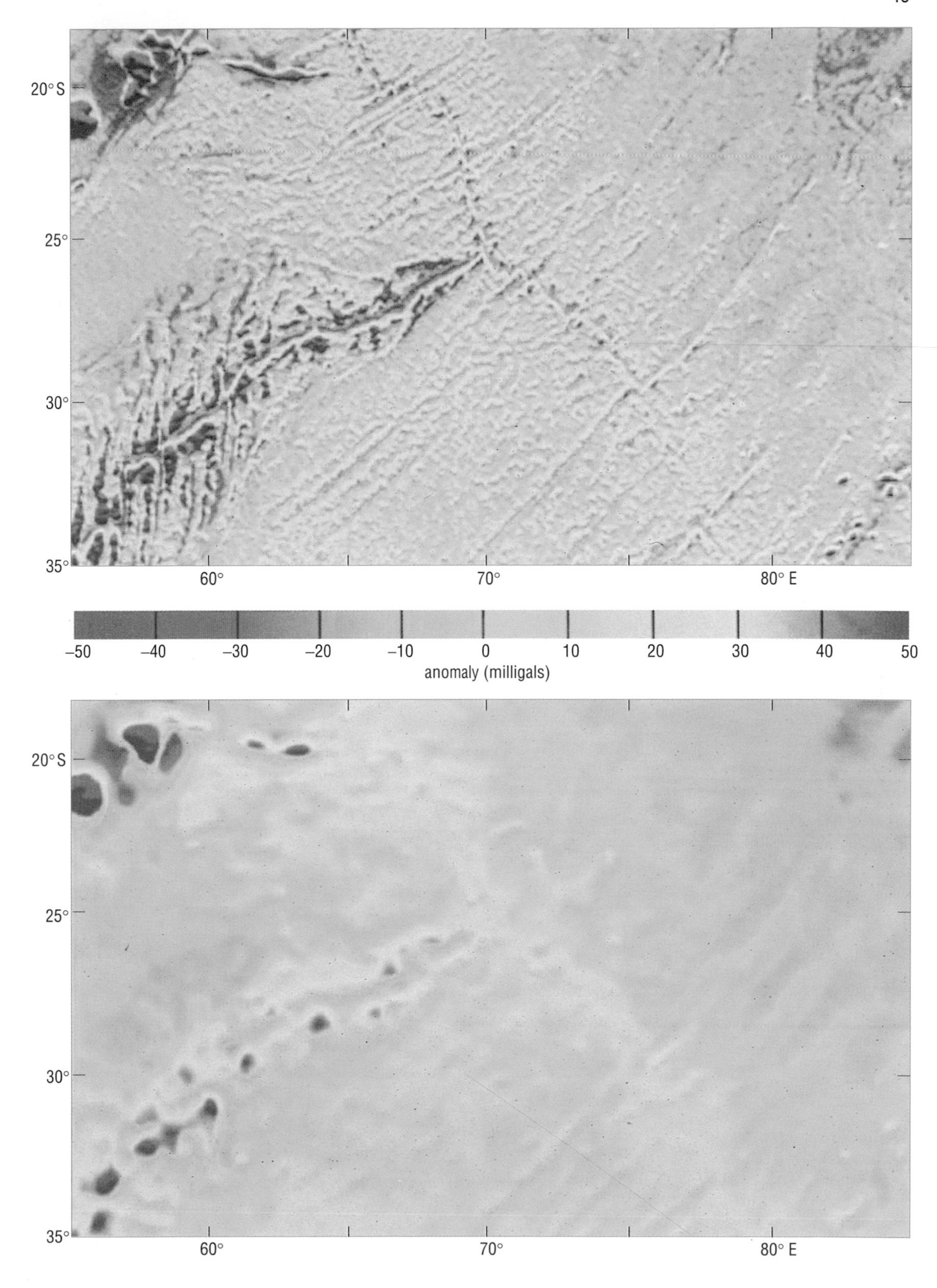

Figure 2.26 Geoid anomaly maps of the same region of the Indian Ocean derived (top) from a combination of *Geosat* and *ERS*-1 data and (bottom) from *Seasat* data. The colour coding represents strength of the geoid anomaly, here expressed in the standard gravity units known as milligals, rather than as heights. Red areas have largest positive anomaly (as demonstrated by locally high sea-surface) and purple areas have largest negative anomaly (as demonstrated by locally low sea-surface).

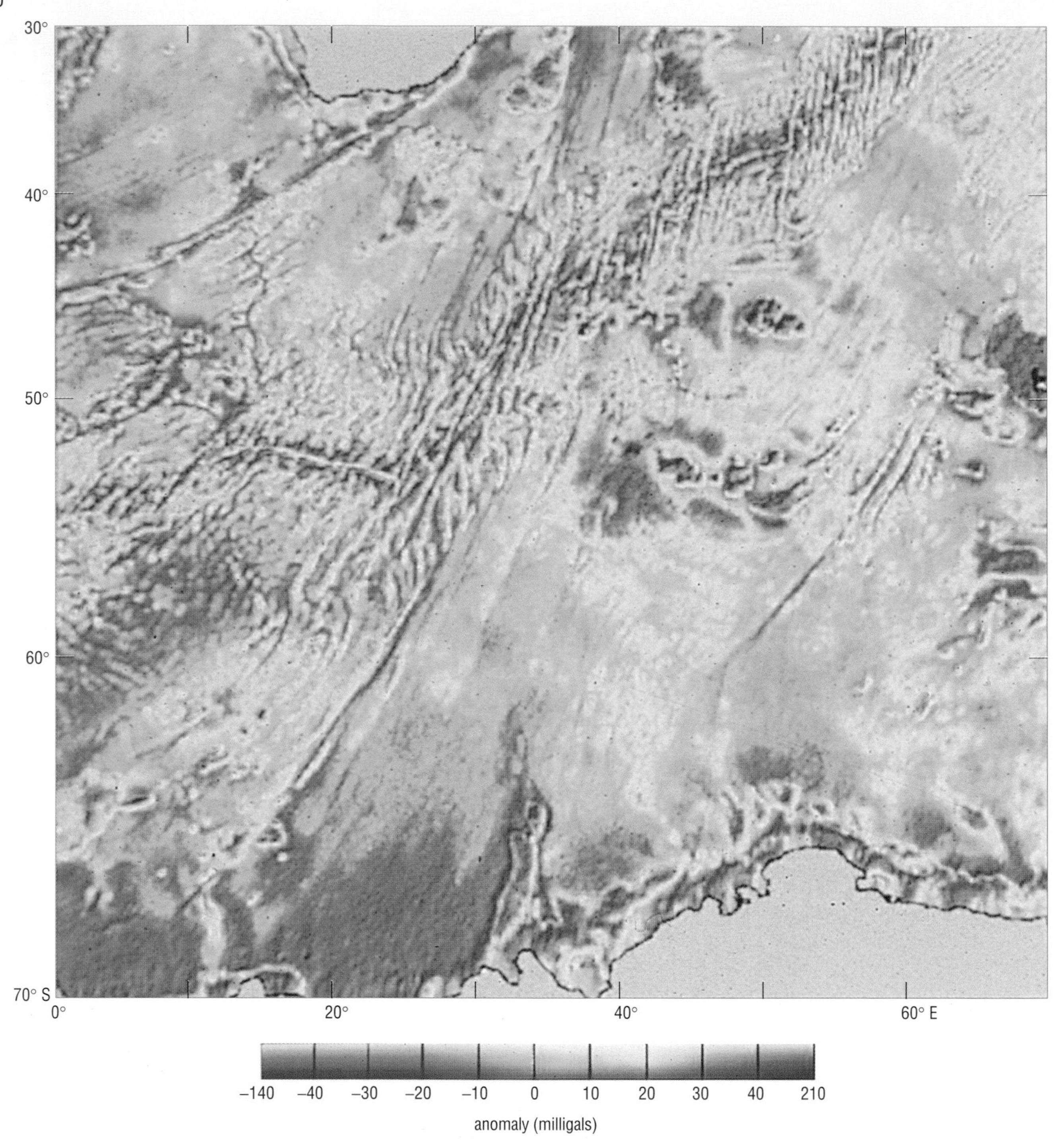

Figure 2.27 (a) Basic geoid anomaly map derived from *Geosat* and *ERS*-1 altimeter data. The area shown extends from South Africa to Antarctica.

The geoid anomaly maps we have looked at so far show either height (in metres) or strength (in milligals) of the geoid anomaly. For many purposes it is more useful if these maps can be converted to show true bathymetry. To do this, artefacts such as those referred to in the answer to Question 2.9(e) have to be corrected, and the scale has to be converted to show depth of the sea-floor. This can be done by calibrating geoid anomaly maps against bathymetric profiles obtained from ships. Provided enough reliable ships' tracks cross the region, the calibrated altimeter-derived bathymetric maps (which show what is referred to as 'predicted bathymetry') are of excellent quality. Figure 2.27(b) is an example taken from the Southern Ocean. Notable differences between the 'predicted bathymetry' and the geoid anomaly map (Figure 2.27(a)) are:

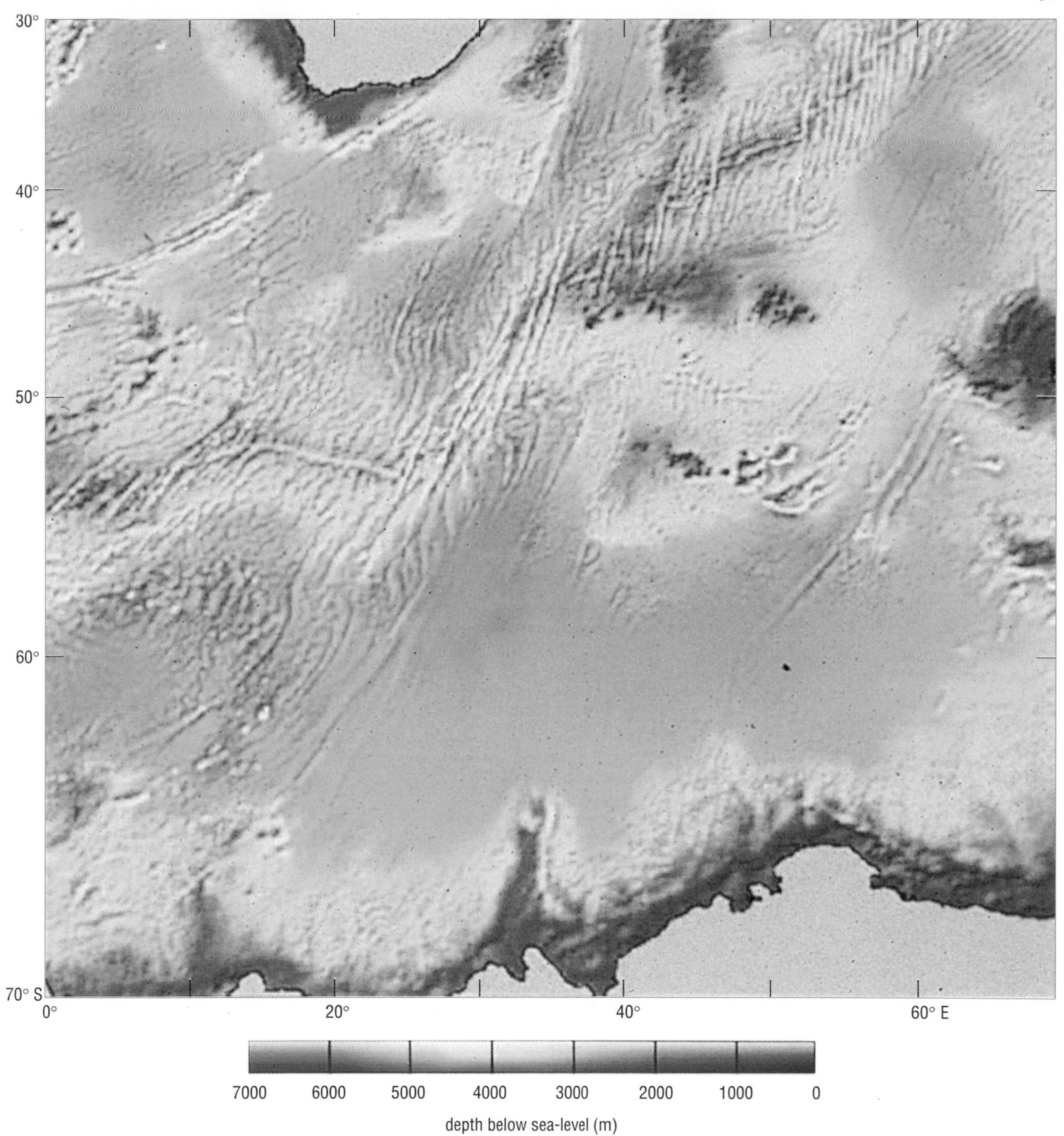

Figure 2.27 (b) The 'predicted bathymetry' obtained by calibrating the geoid anomaly map in (a) against sonar-based bathymetric profiles by survey vessels.

1 the geoid low immediately north of Antarctica between 0° and 30° E does not represent correspondingly deep water (presumably there is a continental shelf here, made of low-density sedimentary rocks);

2 some anomalous geoid lows fringing the Kerguelen and Crozet plateaux (named on Figure 2.24(b)) have disappeared;

3 the Enderby Abyssal Plain is shown smoother than indicated by the geoid anomaly map, presumably because some dense, positive features on the igneous crust are buried beneath lower-density abyssal plain sediments.

Note that Figure 2.27(a) and (b) both reveal more uncharted seamounts in the vicinity of Lena and Marian Dufresne seamounts than Figure 2.24(a).

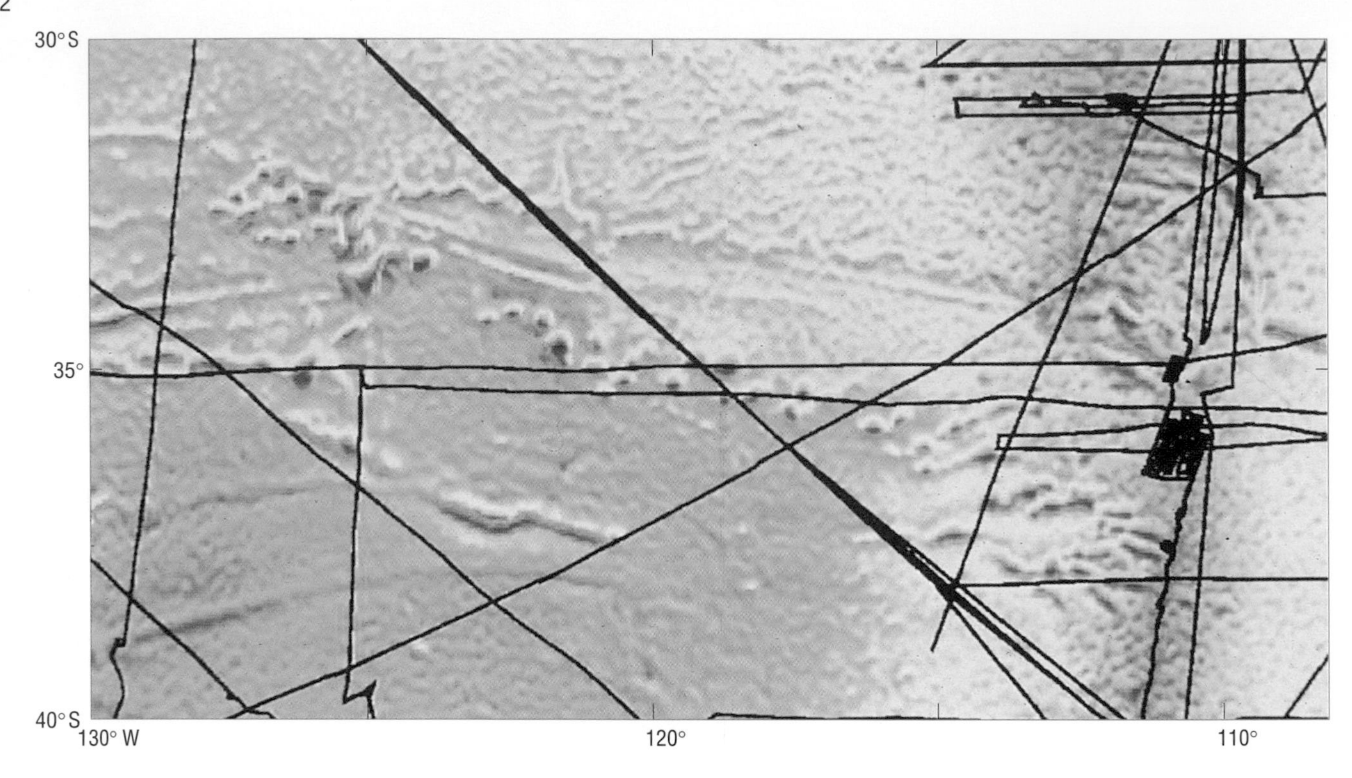

Figure 2.28 Map of 'predicted bathymetry' derived from *Geosat* and *ERS*-1 data, with survey ships' tracks superimposed. The East Pacific Rise runs approximately north–south at about 111° W, and the Foundation Seamounts form a chain stretching west-north-west. (Same colour coding for depth as in Figure 2.27b.)

Satellite bathymetry has proved a very effective way of discovering seamounts throughout the oceans. Figure 2.28 is a map of 'predicted bathymetry' in an area of the southern Pacific that had been relatively poorly surveyed. It reveals a chain of seamounts stretching west-north-west away from the East Pacific Rise, which may have resulted from hot-spot volcanism at a point on the spreading axis. Although this seamount chain was already known (from shipborne surveys and *Seasat* altimetry), the *Geosat/ERS*-1 data show it to contain twice as many seamounts as mapped previously.

2.7 SUMMARY OF CHAPTER 2

1 Aseismic (passive, Atlantic-type) continental margins develop where the continent and adjacent ocean basin belong to the *same* plate. They are underlain by stretched and thinned continental crust, upon which sediments accumulate to build up the continental shelf (covered by shelf seas), slope and rise. Microcontinents are areas of continental crust split off from larger continental masses, which form either islands (crust of near-normal thickness) or submerged plateaux (thinned crust).

2 Seismic (active, Pacific-type) continental margins occur where a continent and the adjacent ocean basin belong to *different* plates, and oceanic lithosphere is being subducted beneath continental lithosphere. These are destructive plate margins. The continental shelf is typically narrower and the slope typically steeper than at aseismic margins, and a trench at the foot of the continental slope generally replaces the continental rise found at aseismic margins. Sediments accumulating in the trenches are partly scraped off to build the inner trench wall oceanwards, and partly subducted. Island arcs occur at another kind of seismic (destructive) margin. These are formed of volcanoes that lie above a subduction zone where one oceanic plate

descends beneath another. They are usually backed by small basins, floored by oceanic crust that has formed by spreading somehow related to the nearby subduction.

3 Ocean ridges (spreading axes) are the most important physiographic (bathymetric) features of the ocean basins. They are the constructive margins of the plates, where new oceanic lithosphere is continually being generated. Gradients on the flanks of slow-spreading ridges are steeper than those on fast-spreading ridges. Median rift valleys are better developed (wider and deeper) on slow-spreading than on fast-spreading ridges.

4 There is a systematic relationship between the depth to the top of the oceanic crust and its age of formation. This is the result of progressive cooling and subsidence of lithosphere with distance from the constructive margin. This relationship enables the age of oceanic crust to be estimated from its depth, and *vice versa*, and works to a good first approximation throughout the major ocean basins.

5 Spreading axes are offset by transform faults that lie along arcs of small circles about the pole of relative rotation of lithospheric plates. Transform faults are seismically active, and they separate plates moving in opposite directions. Beyond the region of offset, they become fracture zones, and lie within a single plate, so seismic activity is much less. These faults and fractures result in scarps and clefts on the ocean floor. Major transform faults are also known as conservative plate margins, but some transform faults can have a component either of spreading (leaky transforms), or of subduction. However, the predominant sense of movement is lateral. Minor offsets (less than about 10 km) of spreading axes are not usually true transform faults, and are described as non-transform offsets.

6 Abyssal plains occupy large areas of deep ocean floor. They are very flat as a result of burial of the rough topography of the oceanic crust by sediments. The sediments are either supplied by turbidity currents from adjacent aseismic margins (e.g. Atlantic Ocean), or deposited from suspension in seawater (pelagic sediment), especially where trenches bordering seismic margins have trapped continent-derived material, and prevented it from reaching the plain (e.g. much of the Pacific Ocean).

7 Seamounts, oceanic islands and aseismic ridges are volcanic features rising from the ocean floor. Linear chains of seamounts and islands, and aseismic ridges, are thought to result from hot-spot volcanism, whereby the oceanic plate moves over an intermittently or continually active fixed source of magma rising from the deep mantle.

8 Satellite altimetry measurements can be used to map features on the sea-floor, because the mean sea-surface height correlates with ocean bathymetry. Conventional bathymetric observations can be refined and extended using these data.

Now try the following questions to consolidate your understanding of this Chapter.

QUESTION 2.12 A seamount 1000 m high has its summit 3000 m below sea-level. What might its age be, and what assumptions do you need to make in arriving at its age? Would you expect the seamount to have a flat top?

QUESTION 2.13 Which of the following statements are true?

(a) The age–depth relationship means that crust generated at a slow-spreading ridge must travel further before subsiding to a given depth than crust from a fast-spreading ridge.

(b) All linear island chains are hot-spot traces.

(c) Sediment thicknesses increase away from ridge crests.

(d) Continental crust below sea-level is thinner than normal.

(e) Sediments in oceanic trenches become subducted.

QUESTION 2.14 In a particular ocean basin a spreading axis (ridge) runs north–south, but it is offset by an east–west transform fault which extends beyond the ridge offset as a fracture zone (Figure 2.29). To the south of the fracture zone, at a distance of 940 km east of the ridge, the depth to the sea-floor is 5000 m. Immediately north of here, on the other side of the fracture zone, the depth is 5500 m. Use the age–depth relationship (Figure 2.13) to answer the following:

(a) Which way is the ridge axis offset – to the left or to the right?

(b) What is the average spreading rate of this ridge?

(c) Is this therefore a 'fast-' or a 'slow-'spreading axis, and which ridges that you have encountered does it resemble?

(d) The position of the ridge north of the transform fault has not been surveyed. Assuming that the spreading rate has remained constant, what is the amount of offset on the ridge due to the transform fault?

(e) How reliable do you think your values are?

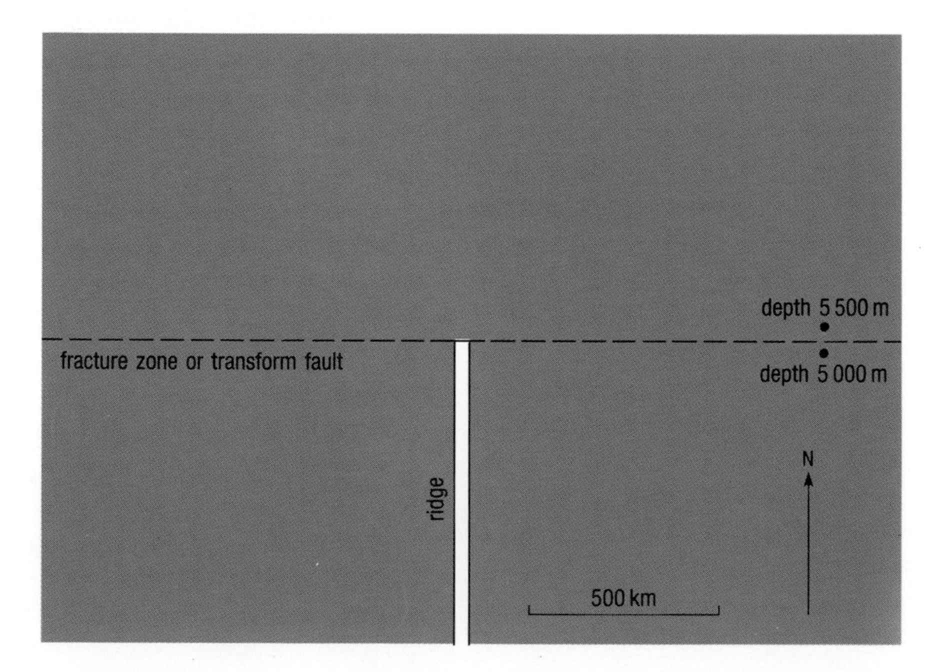

Figure 2.29 Incomplete map of the area discussed in Question 2.14. It will be useful for you to complete the map in answering Question 2.14(d).

QUESTION 2.15 How do you account for the generally smooth appearance of the East Pacific Rise (Figure 2.28), and lack of a discernible median valley on it?

CHAPTER 3 THE EVOLUTION OF OCEAN BASINS

IMPORTANT: In this Chapter you will find frequent reference to the divisions of the geological time-scale. This is shown in the Appendix.

The Earth's oldest rocks – around 3850 Ma old – include both water-lain sediments and evidence of ancient oceanic crust. It follows that oceans have been forming since the beginning of the geological record, and probably before that. However, the shape of most past ocean basins has to be worked out from observations of remnants preserved in continental areas. That is because ocean basins are relatively short-lived features of this planet: no oceanic crust older than about 180 Ma is known from the present oceans.

If we take the life cycle of a large ocean basin to average about 200 Ma, how many times could such basins have been formed since 3800 Ma ago?

The exact answer is 19, but because our figure of 200 Ma is only a crude guess and we do not know how rates of plate-tectonic processes in the distant geological past compared with those of the present, it is wiser to give an approximate figure of 15–20 times. This is probably a minimum, for the Earth's interior was a good deal hotter in the past than it is now, and the turnover of oceanic lithosphere could have been more rapid.

That simple calculation was designed only to give you a feeling for the time-scale of evolution of individual ocean basins, and it is obviously somewhat artificial. In the past, as one ocean basin expanded, another must have contracted, just as the Atlantic and Pacific are doing today. Thus, there is always some overlap in the history of different basins. Continents and ocean basins are continually changing their shapes and relative positions at rates that are geologically very rapid and are not slow even on human time-scales. The speed of sea-floor spreading has been compared with that of growing fingernails. Since the compilation of the first maps to cover any appreciable area of ocean, around five centuries ago, the Atlantic coasts have drawn apart from each other by about 10–20 m. This is a substantial movement, even though it represents only 0.0003% of the width of the ocean. Spreading rates in parts of the Pacific are several times greater than in the Atlantic.

3.1 THE EVOLUTION OF OCEAN BASINS

An individual ocean basin grows from an initial rift, reaches a maximum size, then shrinks and ultimately closes completely. Stages in this cycle are summarized in Table 3.1 (overleaf) and briefly reviewed below.

Whether or not the East African rift valleys really are an incipient ocean basin (Stage 1) and eastern Africa will eventually be split apart is debatable. Nevertheless, such rift valleys must develop along the line of continental separation. When separation does occur, sediments from the adjacent continents soon begin to build out into the new basin and will become part of the eventual continental shelf–slope–rise zone. As the spreading axis migrates away from the marginal areas, the continents become increasingly distant and so the sediment supply dwindles (Stage 2). The ocean floor between the spreading axis and the continent subsides by thermal contraction of the underlying lithosphere (Figure 2.13), abyssal plains form, and the continental shelf–slope–rise zone becomes fully developed. The continental margins are more or less parallel to the central spreading ridge, as in the Atlantic (Stage 3).

Table 3.1 Stages in the evolution of ocean basins, with examples.

Stage	Examples	Dominant motions	Characteristic features
1 embryonic	East African rift valleys	crustal extension and uplift	rift valleys
2 young	Red Sea, Gulf of California	subsidence and spreading	narrow seas with parallel coasts and a central depression
3 mature	Atlantic Ocean	spreading	ocean basin with active mid-ocean ridge
4 declining	Pacific Ocean	spreading and shrinking	ocean basin with active spreading axes; also numerous island arcs and adjacent trenches around margins
5 terminal	Mediterranean Sea	shrinking and uplift	young mountains
6 relict scar	Indus suture in the Himalayas	shrinking and uplift	young mountains

Stage 4 involves the development of one or more destructive plate margins. The reason for the formation of new destructive margins probably lies in changing circumstances in another part of the globe, such as continental collision or the initiation of new continental rifting. If (as seems certain) the Earth is neither expanding nor contracting, the net rates of spreading and subduction over any great circle on the Earth must be equal, and the pattern of plates and plate motion must adjust to keep this so.

The Mediterranean is an ocean in the final stages of its life (Stage 5), with the African Plate being consumed under the European Plate. Unless the world system of plates changes so as to halt the northward movement of Africa relative to Europe, the continental blocks of Europe and Africa will eventually collide, and new mountain ranges will form (Stage 6).

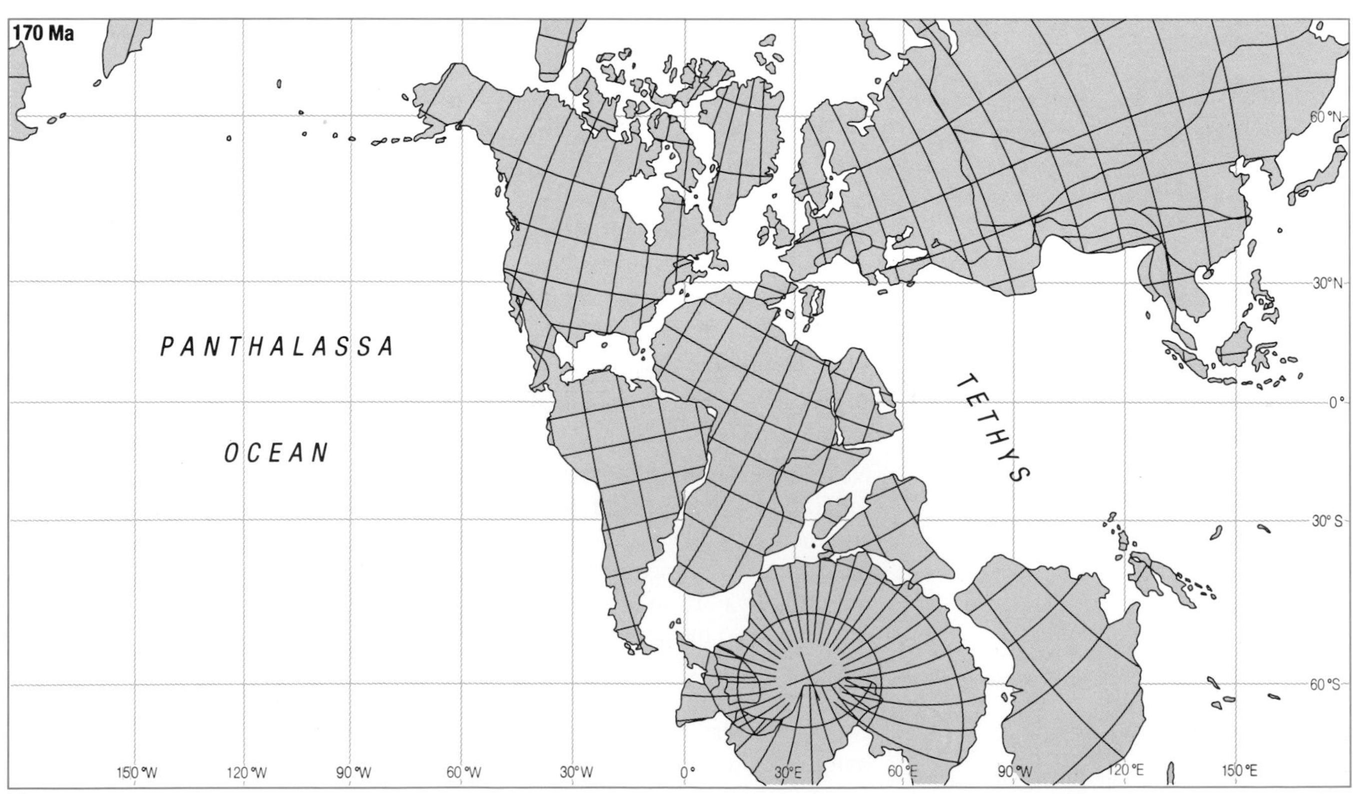

(a)

Figure 3.1 Palaeogeographic reconstruction, compiled from topographic, palaeoclimatic and palaeomagnetic data. Panthalassa was the huge ocean that dominated one hemisphere. Pangea was the supercontinent in the other hemisphere, of which Eurasia and Gondwanaland were two components. (a) Jurassic, about 170 Ma ago. (b) Cretaceous, about 100 Ma ago. (c) Eocene, about 50 Ma ago. The maps show *present-day* coastlines for ease of reference. Ancient coastlines did not coincide with these.

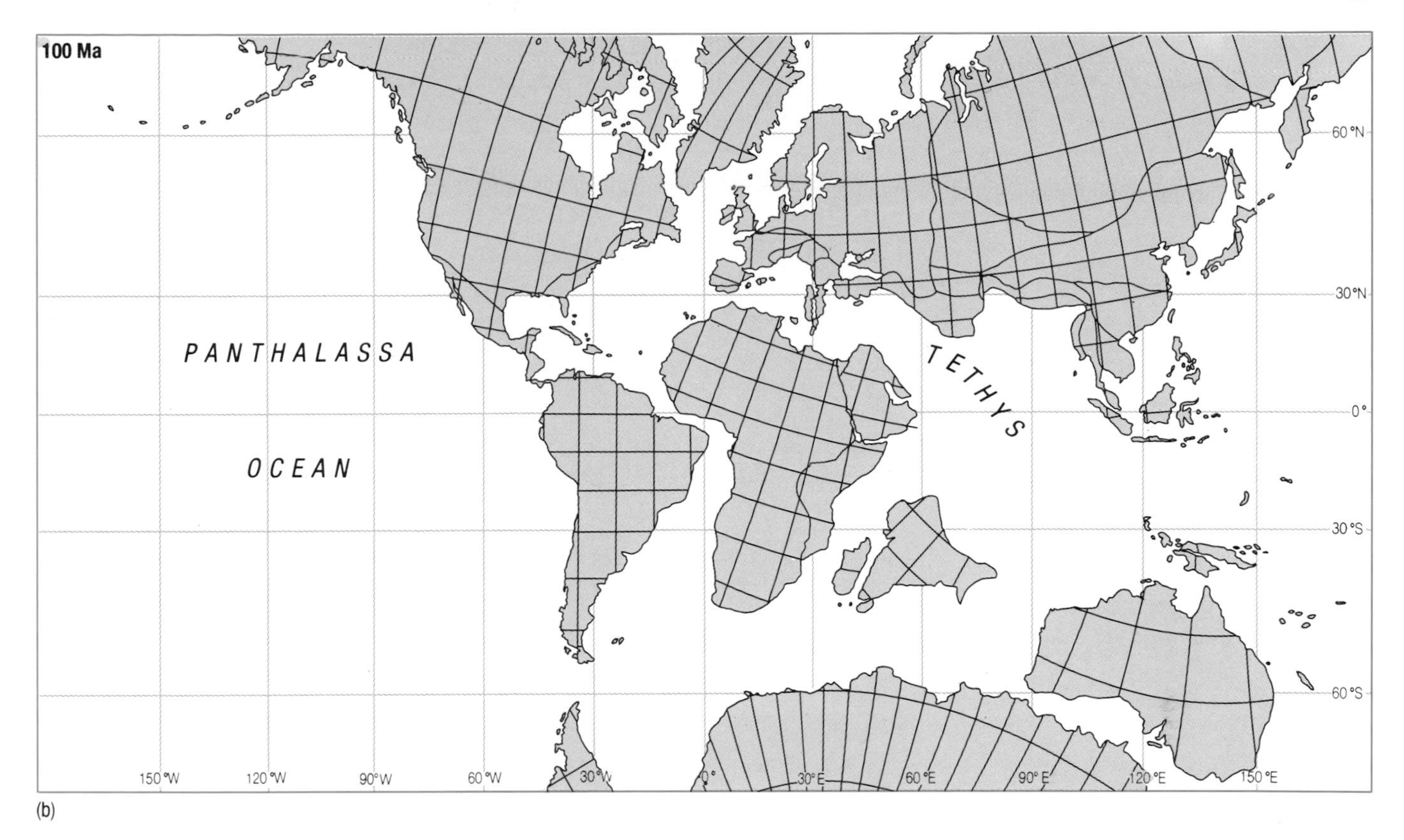
100 Ma
PANTHALASSA
OCEAN
TETHYS
60 °N
30 °N
0°
30 °S
60 °S
150 °W
120 °W
90°W
60 °W
30°W
0°
30°E
60 °E
90° E
120 °E
150 °E

(b)

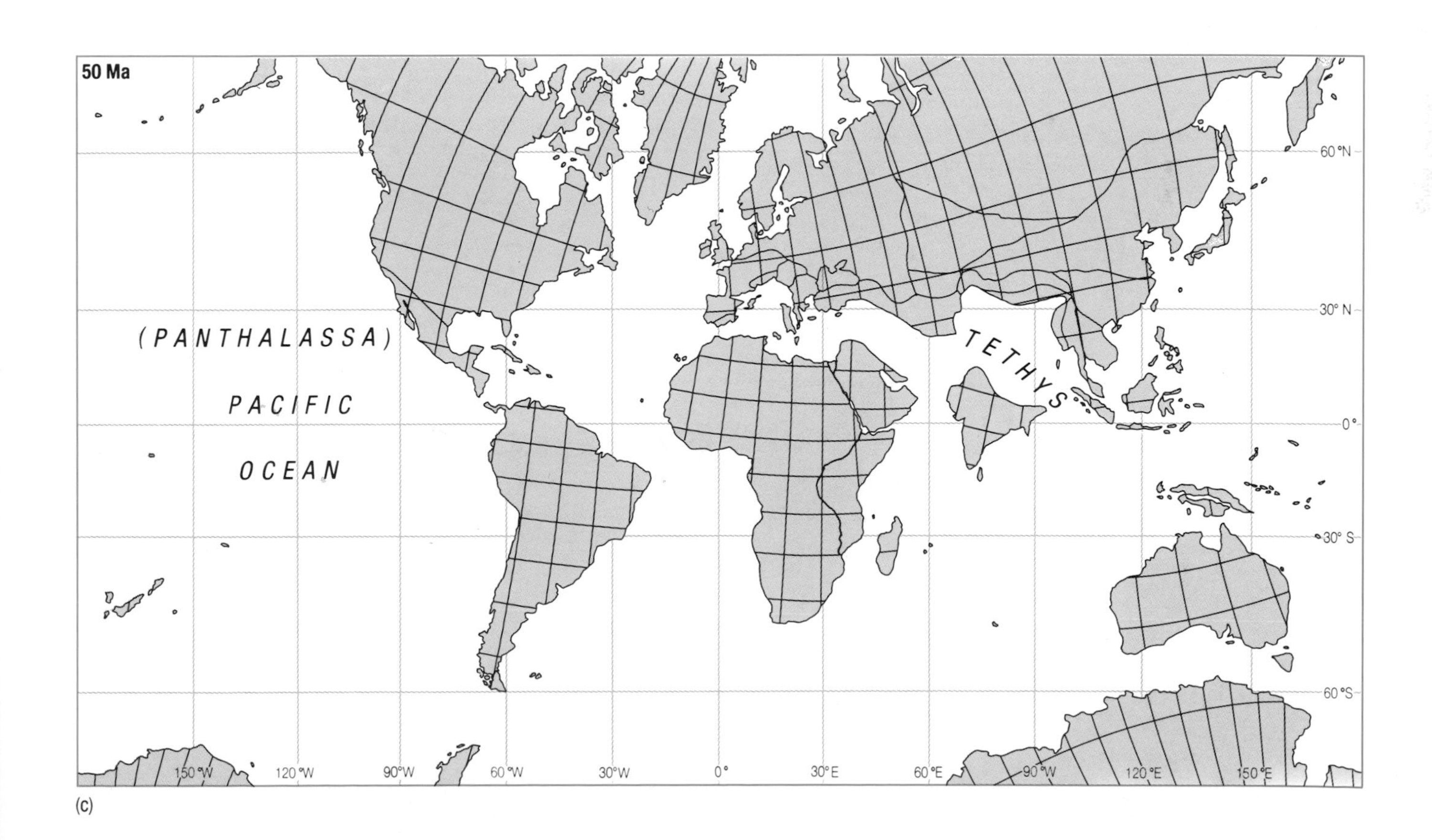
50 Ma
(PANTHALASSA)
PACIFIC
OCEAN
TETHYS
60 °N
30° N
0°
30° S
60 °S
150 °W
120 °W
90°W
60 °W
30°W
0°
30°E
60 °E
90 °W
120 °E
150 °E

(c)

QUESTION 3.1 Study Figure 3.1.

(a) Tethys was a major ocean basin, a branch of Panthalassa, that once separated Eurasia from the southern continent, Gondwanaland. How have its shape and size changed over the past 170 Ma and what are its present-day marine remnants?

(b) How long ago did the Atlantic Ocean start to open?

(c) When did the Indian Ocean start to open?

(d) Was the period from 170 to 100 Ma or the one from 100 to 50 Ma more significant in terms of the fragmentation of Gondwanaland?

(e) What were the major changes between 50 Ma and the present day?

3.2 THE BIRTH OF AN OCEAN

Figure 3.2 summarizes the development of a new ocean basin. During crustal extension, the ductile lower part of the crust is stretched, but the brittle upper

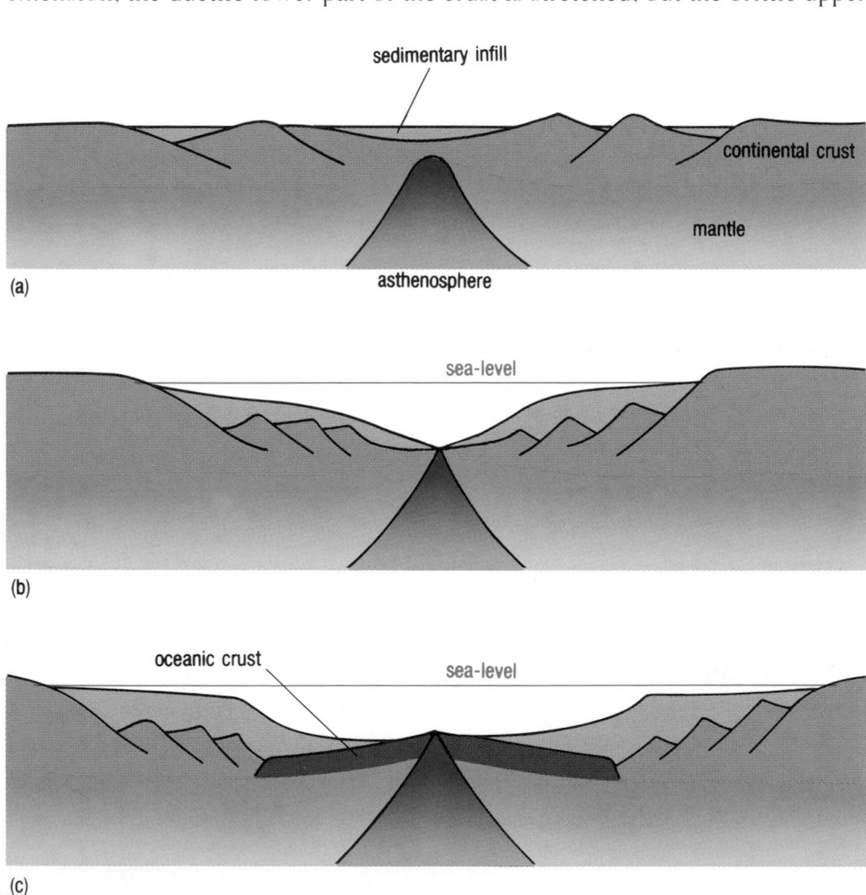

Figure 3.2 Diagrams illustrating how a new ocean basin may form.
(a) The surface of the stretched and rifted region is still above sea-level and may even be uplifted as a result of thermal expansion because of a heat source at the site of the future spreading axis. Terrigenous sediments (i.e. those derived from erosion on land) occupy the rift valley.
(b) The future continental margins have thinned enough to subside below sea-level and marine sedimentation has begun. Sediments thicken away from the new spreading axis.
(c) Separation is complete, a new spreading ridge has developed, and a shelf–slope–rise zone is forming (cf. Figure 2.6).

part is rifted. Blocks of crust slide down fault planes, and sediments accumulate in the lakes and valleys which occupy the resulting depressions. When separation occurs, basaltic magma rises to fill the gap between the two continental blocks. Because the resulting new oceanic crust is both thinner and denser than continental crust, it lies below sea-level. The remainder of the lithosphere, below the crust, is composed of upper mantle material.

Initially, the young marine basin is fairly shallow. If repeated influxes of seawater become wholly or partly evaporated, salt deposits (**evaporites**) will accumulate. Otherwise, there will be normal marine sedimentation of muds, sands and limestones, depending on local conditions. One of the clearest examples of a young ocean basin is the Red Sea.

3.2.1 THE RED SEA

In the Red Sea, a narrow deep axial zone is flanked on either side by a broad shallow area of shelf sea (Figure 3.3(a)). Evaporites of Miocene age (deposited between about 20 and 5 Ma ago) that are over 4 km thick in places underlie the shallower waters of these flanking regions. They obscure the nature of the crust beneath, which appears to be thinned and stretched continental crust (Figure 3.3(b)).

The evaporites were deposited at a time when the only marine connection to the Red Sea was with the Mediterranean by an intermittent, shallow, seaway. Evaporite deposition ended about 5 Ma ago at the end of the Miocene when this seaway was finally broken and a new connection with the Indian Ocean was opened in the south. Open water conditions were established, in which planktonic organisms flourished, especially in the southern Red Sea. High rates of biogenic (biologically derived) sedimentation caused bathymetric features to be smothered, and they become much less obvious south of about 16° N.

Further north, the post-Miocene biogenic sediments give way to a thinner sequence of terrigenous (land-derived) clays, sands and gravels, produced by erosion of the flanks of the basin. Similar terrigenous sediments can also be found interbedded with the Miocene evaporites, especially near the margins.

Only in the axial zone, which represents that part of the Red Sea generated since the end of evaporite deposition, do we find true oceanic crust produced by sea-floor spreading. On the basis of seismic and magnetic surveys, submersible observations and side-scan sonar mapping, the axial zone can be subdivided into several regions along its length (Figure 3.3(a)) as described below.

Figure 3.3 (a) Outline map of the Red Sea, with the axial zone (dark blue) defined by the 500-fathom isobath and subdivided into four main sections, described in the text (1 fathom = 6 feet = 1.83 m).
(b) Highly schematic cross-section through the Red Sea.

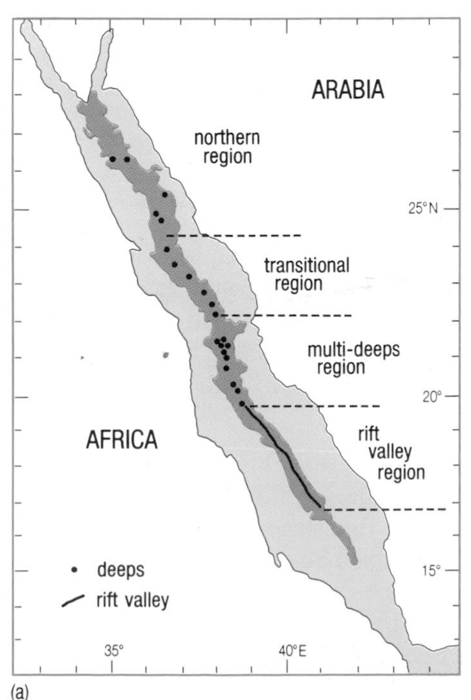

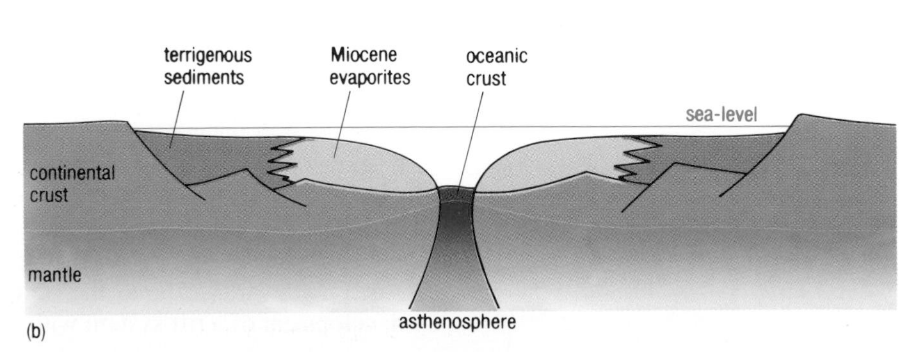

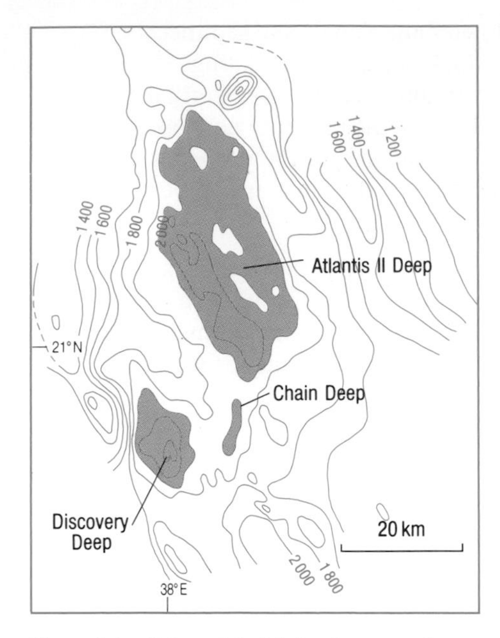

Figure 3.4 Bathymetric details of some major 'deeps' in the multi-deeps region of Figure 3.3(a). Hot, metal-rich brines are found in them, and metalliferous muds are being deposited there. Depths are given in metres.

Rift valley region

The southern part of the axial trough is now known to have a well-developed straight central rift (similar to that on the Mid-Atlantic Ridge; Figure 2.12), which is offset by 3–10 km about every 30–50 km. These discontinuities may be either transform faults or some sort of non-transform offset.

High-amplitude linear magnetic anomalies occur throughout this region, though they become weak and irregular at the offsets. Measurements of the magnetic anomaly stripes indicate that spreading has proceeded at a rate of about 0.8 cm yr^{-1} for the past 5 Ma.

Multi-deeps region

North of about 20° N, the straight axial rift loses its identity and is replaced by a complex series of axial deeps, distributed partly in an *en echelon* fashion, perhaps because of offsets by transform faults. The deeps are best developed between about 20° N and 22° N and they have attracted commercial interest on account of the metal-rich hot brines and muds which some of them contain (Figure 3.4). Individual deeps have a rift-valley type structure with strong magnetic anomalies, but between the deeps the anomalies are much weaker and the axial region is sediment-covered.

Transitional and northern regions

Beyond about 22° N, the deeps become progressively narrower and less well developed, and the associated magnetic anomalies suggest that the oceanic crust in them may be only 2 Ma old or less. North of about 25° N, only isolated deeps are found, the high-amplitude magnetic anomalies characteristic of oceanic crust which occur further south have virtually disappeared, and the region appears to have a more or less continuous sediment cover.

In summary, then, only about 80 km width ($2 \times 5 \times 10^6 \times 0.8 \times 10^{-2}$ m) of new ocean floor up to 5 Ma old can be demonstrated to have formed in the southern part of the axial zone; further north, ocean floor has formed only in the deeps and is 2 Ma old or less.

All of this suggests strongly that the axial zone of the Red Sea is a northward-propagating zone of separation between adjacent plates of continental lithosphere. The fracture began to open properly about 5 Ma ago in the south but has yet to do so in the north. This is consistent with the end of evaporite deposition in the Red Sea about 5 Ma ago, when a link with the Indian Ocean was established via the Gulf of Aden.

But what about the crust outside the axial region of the Red Sea? Figure 3.5 shows magnetic anomalies in the southern Red Sea. The well-defined axial anomalies that are characteristic of true oceanic crust give way to an irregular and much weaker pattern in the flanking regions. This is consistent with thinned and stretched continental crust, injected by thin vertical sheets of basaltic rock (dykes).

From a variety of geological evidence, it is clear that this stretching and subsidence of the continental crust pre-dates the sea-floor spreading in what became the Red Sea. It seems likely that the first stage, about 35 Ma ago, was the propagation of a crack from the Arabian Sea westward through the Gulf of Aden. By about 25 Ma ago, east–west extension had begun to be felt across the entire area of the Red Sea (from its junction with the Gulf of Aden northwards as far as the Gulf of Suez), and was manifested by the development of a rift system within the continental crust. A new fracture

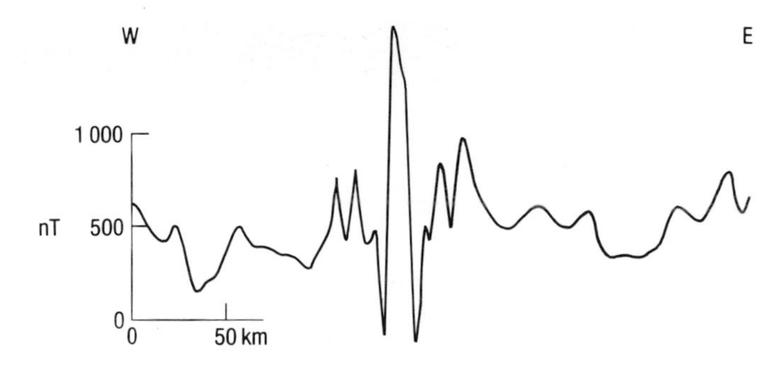

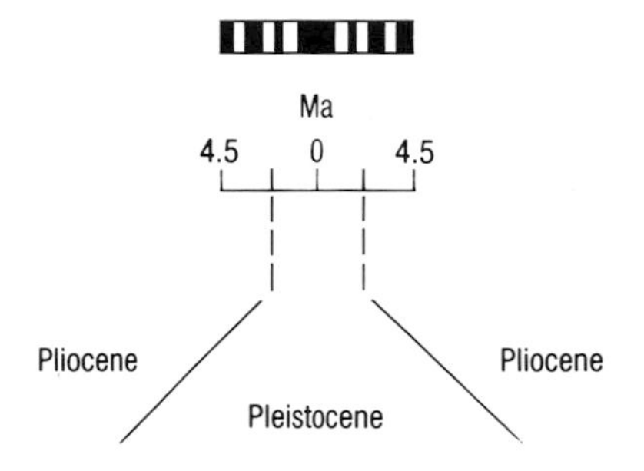

Figure 3.5 Magnetic anomalies for the southern Red Sea, showing the contrast between the strong central pattern over the axial trough and the subdued pattern on either side. For this profile, correlation with the magnetic reversal time-scale (black = normal polarity; white = reverse polarity) gives an average full spreading rate of 1.5 cm yr^{-1} over the past 4.5 Ma. The magnetic field is measured in nT (nT = nanotesla = 10^{-9} T).

developed in the north along what is now the Gulf of Aqaba/Dead Sea line, running towards the north-north-east. Transcurrent (lateral) movement along this line accompanied further widening of both the Gulf of Aden and the Red Sea as Arabia moved away from Africa. Oceanic crust formed in the Gulf of Aden. About this time there was widespread extension-related volcanism in the western parts of what are now the Yemen, Saudi Arabia, Eritrea, and northern Ethiopia.

Along the Red Sea rift, the continental crust continued to stretch and subside, and evaporites were deposited on top of it. About 5 Ma ago the system was reactivated when movement was renewed along the Gulf of Aqaba/Dead Sea line. The continental crust was already so fully stretched that rather than thinning and stretching even further it was pulled apart, and sea-floor spreading began here for the first time, while continuing in the Gulf of Aden.

The opening of the Red Sea as a new ocean was clearly a drawn-out and complicated affair, and we should bear in mind that the opening stages of other ocean basins are likely to have been similarly complex. However, we must now move on to consider larger ocean basins that have reached Stage 3 or 4 of Table 3.1, bearing in mind that in the early stages of formation, they probably resembled the Red Sea.

QUESTION 3.2 Roughly where would you expect to find the pole of relative rotation (see Figure 2.14) about which the African and Arabian Plates are moving to form the Red Sea?

3.3 THE MAJOR OCEAN BASINS

Figure 3.6 summarizes the age distribution of oceanic crust beneath the world's oceans, as determined from magnetic anomaly patterns. The virtually symmetrical pattern of ages about the ocean ridges is visible everywhere, and it is clear that (apart from the Caribbean area and in the extreme south-west) the Atlantic has had the least complicated evolution of any of the three main ocean basins. The Pacific and Indian Oceans display more complex histories, partly because of the development of major subduction zones along one or more boundaries and partly because of adjustments in spreading direction.

QUESTION 3.3 With reference to Figure 3.6, determine the relative order in which sea-floor spreading began in these three parts of the Atlantic:
(i) northernmost Atlantic, between Greenland and north-west Europe;
(ii) northern Atlantic, between North America and north-west Africa;
(iii) southern Atlantic, between South America and southern Africa.

From Figure 3.6, it is relatively easy to reconstruct stages in disruption of the continental jigsaw puzzle that led to the present-day Atlantic. It is simply a matter of moving the continents back along the transform faults to determine the positions of their margins at any particular time, as represented by the age of the ocean floor magnetic stripes.

The Pacific and Indian Oceans are more difficult to 'close up'. The Pacific is almost surrounded by subduction zones, so much of the evidence of its older history has disappeared. The widths of the ocean floor age strips increase northwards along the line of the East Pacific Rise in a way that is consistent with increasing spreading rates from south to north. However, the East Pacific Rise runs into the North American continent, beneath which its northern end is being subducted.

You have already seen evidence of changes in spreading direction in the north-western Pacific (Figure 2.22), and more evidence is to be seen in the pattern of ocean floor ages in Figure 3.6.

The Hawaiian Chain is oblique to the spreading trend indicated by the age strips for 0–43 Ma Pacific Ocean floor in Figure 3.6. What does that tell you about the movement of the East Pacific Rise? Bear in mind that hot spots can be regarded as effectively stationary with respect to the Earth as a whole.

The East Pacific Rise must have been moving relative to the frame of reference provided by the Hawaiian and similar chains (the 'hot-spot reference frame'). It is important for you to realize that magnetic anomalies, and the sea-floor age strips which can be mapped using them, record only the motion *relative to the spreading axis at which the sea-floor was formed*. It is quite possible for the spreading axis itself to be migrating relative to the deep Earth. In fact, it is impossible to have several spreading axes on a sphere and keep them all stationary, unless the plate motion at each one is individually and exactly compensated for by nearby destructive plate margins. Thus, a spreading axis can migrate across an ocean basin, and will be destroyed if it gets carried into a subduction zone. Figure 3.7 (overleaf) shows how the global plate boundaries are thought to have evolved over the past 61 Ma based on the types of data you have read about in this and the previous Chapter, especially hot-spot traces and magnetic anomaly patterns.

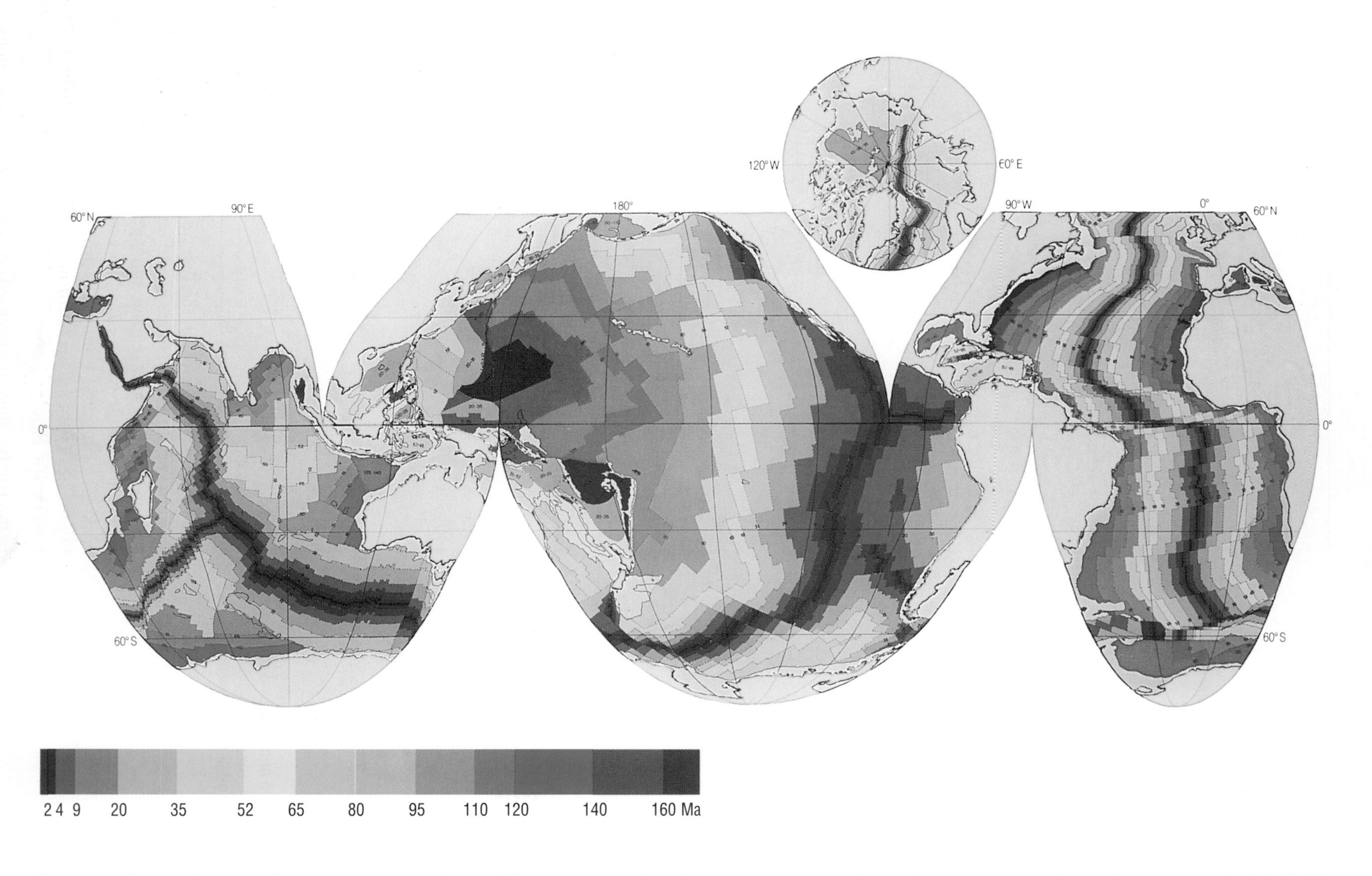

Figure 3.6 The age of the ocean floor, showing strips of floor of different ages derived mainly from measurements of magnetic anomaly stripes. Boundaries are drawn at 2, 4, 9, 20, 35, 52, 65, 80, 95, 110, 120, 140 and 160 Ma intervals in a colour scheme that runs from dark grey (youngest) through red, yellow, and green to blue (oldest). Pale brown areas are the continental shelves.

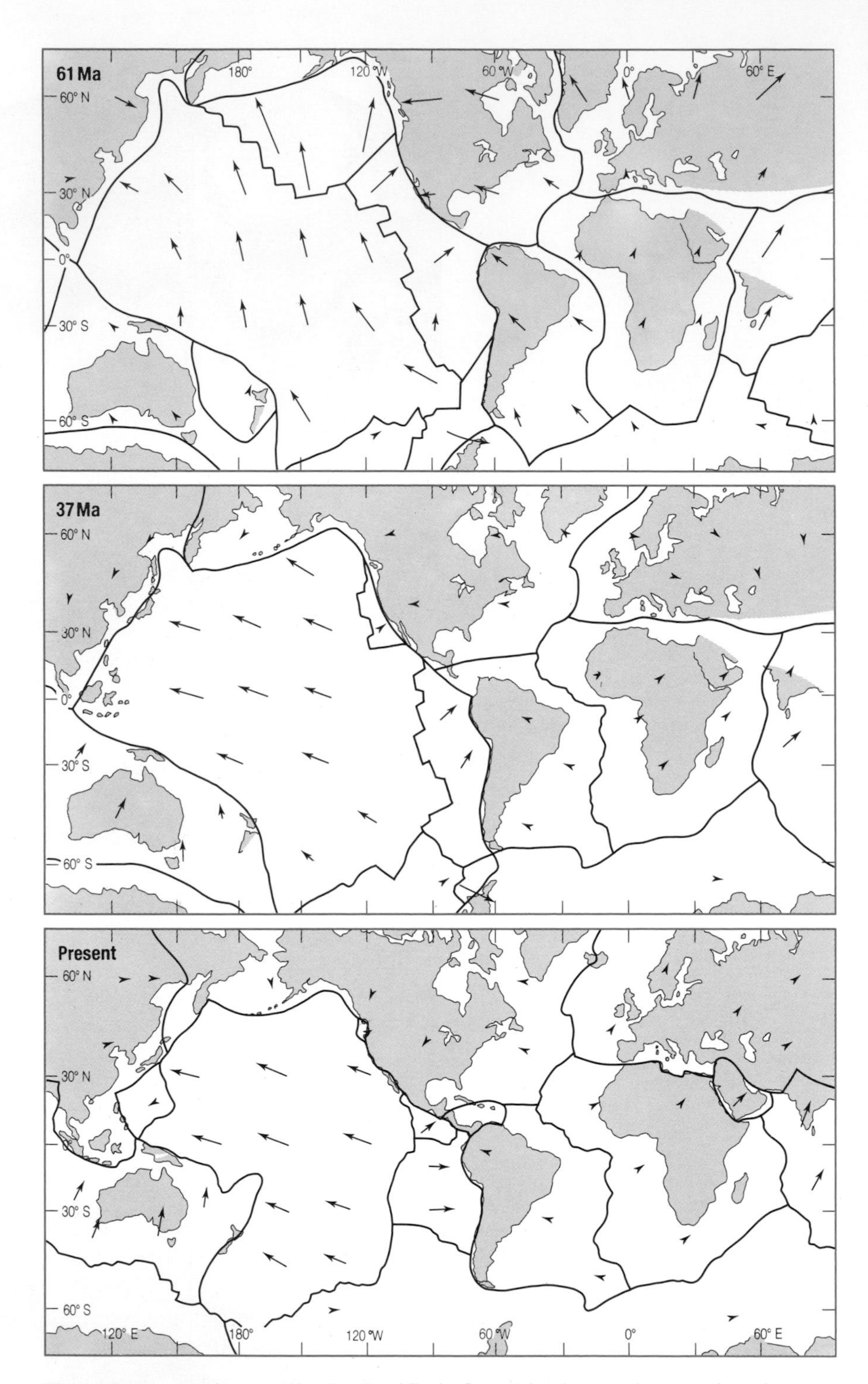

Figure 3.7 Evolution of plate boundaries over the past 61 Ma. The arrows show the directions and relative rates of motion of the plates.

The oldest oceanic crust in the Pacific is found in the north-west, but the western Pacific as a whole is an area of great complexity. This is because of the generation of new oceanic lithosphere at various spreading axes above subduction zones, where island arcs are being built and then split apart, and back arc basins are forming, as outlined in Section 2.2.2 and illustrated in Figure 3.8. These events occur independently of sea-floor generation at the East Pacific Rise.

In the Pacific, there is an added complication when it comes to reconstruction of the continental jigsaw puzzle. Palaeomagnetic, geological

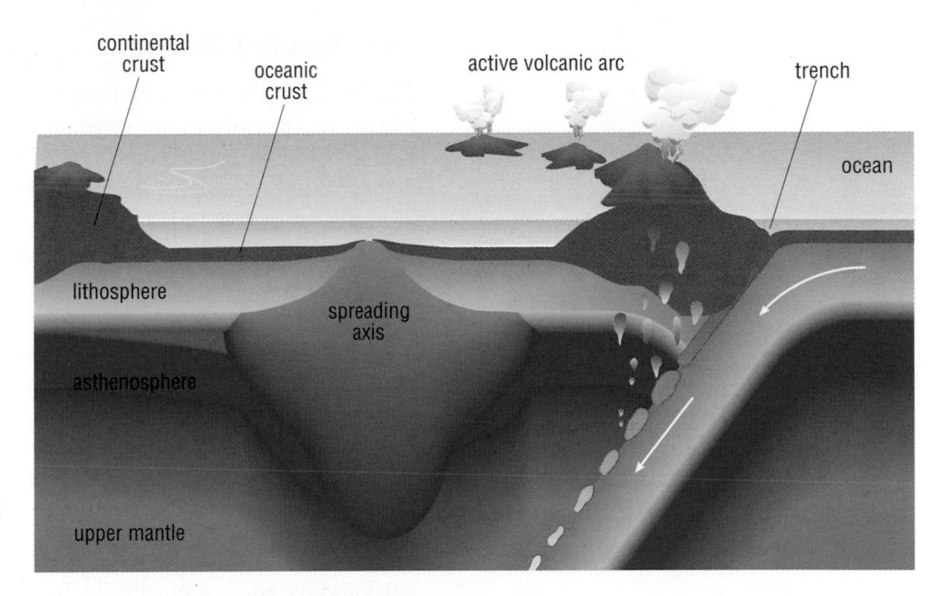

Figure 3.8 A schematic cross-section through a typical western Pacific continental margin. A small spreading axis has generated new oceanic crust behind the island arc. In some places, an even more complex series of island arcs and back arc basins separates the deep ocean trench from the continental margin.

and palaeontological evidence have demonstrated that substantial tracts of western North America are 'exotic terranes'. These are blocks of continental crust – microcontinents – that have been transported by plate movements for great distances across the Pacific to become accreted onto North America.

The Indian Ocean has several features of interest, too. Its northern boundary is a major complex subduction zone, represented by the Himalayan belt and the Java Trench system. South of India, the spreading direction changed from north–south to north-east–south-west about 50 Ma ago when the present South-east Indian Ridge became established.

The aseismic Ninety-east Ridge (Section 2.5.4) must lie on the line of an old major transform fault, for the age of the crust changes in opposite directions on either side of it, as can be seen in Figure 3.6.

QUESTION 3.4

(a) In what directions do crustal ages change on either side of the Ninety-east Ridge?

(b) Were these areas of oceanic crust generated before or after development of the present South-east Indian Ridge?

(c) What happened to the spreading axis which generated the crust immediately east of the Ninety-east Ridge?

You can find other examples of changes of spreading rate and direction, and development of new spreading axes and subduction zones, displayed in Figure 3.6. You may also have noticed that many of the ocean floor age strips are oblique to subduction zones, e.g. in the north-east Indian Ocean and parts of the western Pacific. Oblique subduction of ocean floor is by no means exceptional, and it means that continent–continent or continent–island arc collision does not necessarily occur head-on, and so major transcurrent faulting may result.

In the major ocean basins, irrespective of whether they are classified as Stage 3 or 4 in Table 3.1, there is no indication that increasing age is correlated with any decline in the intensity of sea-floor spreading activity. The Pacific basin is the oldest, for instance, but it has the fastest spreading rates. When we come to Stage 5, we find that even in the latest stages of evolution, there is little diminution of vigour.

3.3.1 THE MEDITERRANEAN

The Mediterranean can be classified as an ocean in the final stages of its life cycle, the only major remnant of the once-extensive Tethys Ocean (Table 3.1, Stage 5; Figure 3.1). The Mediterranean is shrinking as the African Plate continues to thrust its way northwards beneath the European Plate. We might therefore expect that the Mediterranean would be floored by oceanic crust dating back perhaps as far as Jurassic times, which would be consistent with its being the remnant of an old ocean, and that it would have an obvious major trench. It turns out, however, that the Mediterranean region has been broken into many small plates, whose boundaries may be delineated partly by analysis of earthquakes (which are sporadic and scattered in this region), and partly by drilling. The deep basins contain several kilometres of sediments, including evaporites (see Chapter 6), and this hampers geophysical investigations into the nature of the underlying crustal layers.

The eastern Mediterranean is floored by crust of Cretaceous age (*c.* 110 Ma), bordered by an area of middle Tertiary (25 Ma) crust south of Italy, possibly the result of back-arc spreading. There is a collision zone running south of Cyprus, where the African Plate meets the Eurasian Plate, and there is some evidence of subduction there. Miocene and younger (10–2 Ma) oceanic crust in the western Mediterranean is generally believed to have formed in a back-arc setting, associated with the subduction that removed older ocean floor from this region. Active volcanism and frequent earthquakes in and around the Mediterranean show that this ocean basin is still evolving.

QUESTION 3.5 Can you explain the correlation between crustal age in the eastern and western Mediterranean (Figure 3.6) and the geoid anomaly in these two areas (Figure 1.18)?

Apart from the eastern Mediterranean sea-floor, ocean crust older than about 70 Ma is represented in the Mediterranean region by slivers of oceanic crust and upper mantle which have been tectonically removed from the ocean floor and emplaced over a continental margin, usually during a collision event. A piece of ocean floor which has been preserved in this way is called an **ophiolite** or **ophiolite complex**. Ophiolites are too small and fragmentary to preserve magnetic stripes, but they can be dated by other methods (e.g. radiometrically or by examining the fossils in the oldest sediments associated with them). The ophiolite forming the Troodos mountains in Cyprus is one of the most intact and best-known. This is one of many ophiolites of mostly late Cretaceous age (*c.* 80 Ma), preserved in collisional mountain chains along the north side of the Mediterranean in the Balkans and Asia Minor. This belt runs eastwards all the way through the Himalayas and represents the remains of the Tethys Ocean. Older ophiolites tend to occur near the western Mediterranean, particularly in the Alps, and range in age back to the Triassic (*c.* 220 Ma). These may represent the oldest ocean floor of the Mediterranean–Tethys region.

Reconstruction of the evolution of ocean geometry is more than just an academic exercise. Combined with information from the sediments on the ocean floor, it can be used to reconstruct past climatic and circulation patterns. Compilations such as Figure 3.6 are continually being up-dated as less well-known areas are surveyed in greater detail, sometimes with the help of satellite bathymetry data. The more we understand about fluctuations in past oceanic cycles, the better we shall understand the present-day oceans.

We shall look at this with particular reference to sea-level in Chapter 6, but next we turn to the mechanisms of generation of new oceanic crust.

3.4 SUMMARY OF CHAPTER 3

1 Oceanic crust is much younger than most continental crust. Oceanic lithosphere must have been generated at spreading axes (ridges) and destroyed at subduction zones many times since the formation of the Earth. Ocean basins form initially by stretching and splitting (rifting) of continental crust, and the rise of mantle material and magma into the crack to form new oceanic lithosphere.

2 The Red Sea is an embryonic ocean that appears to be opening progressively from the south, where the axial region is underlain by oceanic crust and has a rift valley. Further north are isolated deeps – with metal-rich muds – but there is less evidence of oceanic crust in the axial region. The thick evaporites bordering the axial region rest on thinned continental crust.

3 Among the major ocean basins, the Atlantic has the simplest pattern of ocean-floor ages. Subduction is confined to relatively small island arc systems in the Caribbean and the extreme south-west. Successive stages in the shape of the Atlantic basin are therefore fairly easy to reconstruct, by moving the continents back along a direction at 90° to the magnetic anomaly stripes and parallel to the transform faults.

4 In contrast, both the Pacific and Indian Oceans (which have major subduction zones) are characterized by changes of spreading rate and direction and the development of new spreading axes. Because of these complications, it is difficult to work out how the shapes of these ocean basins have changed with time. The occurrence in western North America of 'exotic terranes', which in some cases are believed to have originated as microcontinents in the south-west Pacific, make the task of such reconstruction even more complicated.

5 The Mediterranean represents an ocean in the final stages of its life cycle, contracting as Africa pushes northwards into Europe and western Asia. However, no *in situ* oceanic crust older than the Cretaceous is known from the Mediterranean basin, and crust as young as 2 Ma has been found there, demonstrating that there is no inverse correlation between age of an ocean basin and the vigour of sea-floor spreading and plate-tectonic activity.

Now try the following questions to consolidate your understanding of this Chapter.

QUESTION 3.6 'India moved northwards at rates of between 10 and 20 cm yr^{-1} from about 135 to about 45 Ma ago. For the next 25 Ma or so, it moved more slowly at about 5 cm yr^{-1}.' Are the relative widths of the age bands south of India in Figure 3.6 consistent with these statements?

QUESTION 3.7 According to Figure 3.6, what is the approximate ratio between the average spreading rate over the past 52 Ma for the East Pacific Rise at the Equator and that of the Mid-Atlantic Ridge at 30° S over the same time (52 Ma is in the middle of the yellow field in Figure 3.6)? If the average spreading rate in the Atlantic was 2 cm yr^{-1} over this period, what was it for the East Pacific Rise?

CHAPTER 4 THE STRUCTURE AND FORMATION OF OCEANIC LITHOSPHERE

Knowledge of the nature of the oceanic crust and upper mantle comes from four main sources: (1) geophysical techniques, notably seismic refraction and reflection, but also magnetic and gravity surveys and heat flow measurements; (2) examination and measurement of physical properties of rocks dredged from the sea-bed and cored from the upper parts of the crust; (3) direct observation and photography of the sea-bed using submersibles; and (4) land-based studies of ophiolites. It was seismic refraction studies which first revealed that the oceanic lithosphere consists of layers whose seismic velocity (the speed at which sound waves travel through the rock) increases with depth. These **seismic layers** are numbered from 1 (at the top) to 4, and are usually matched with rock types according to the scheme shown in Figure 4.1.

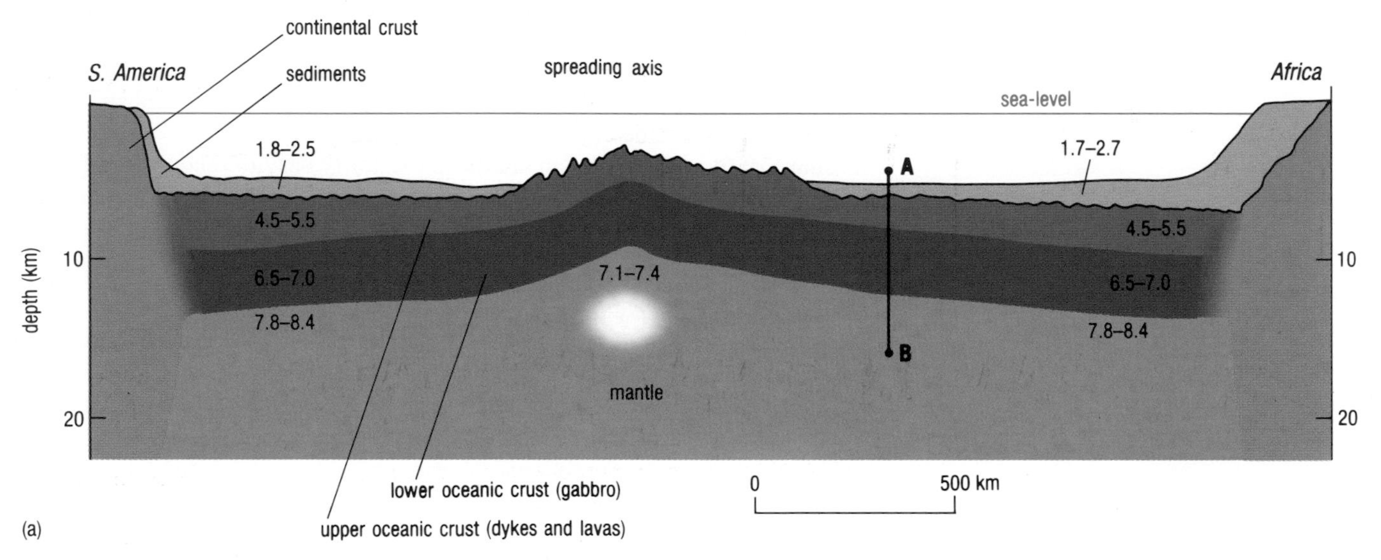

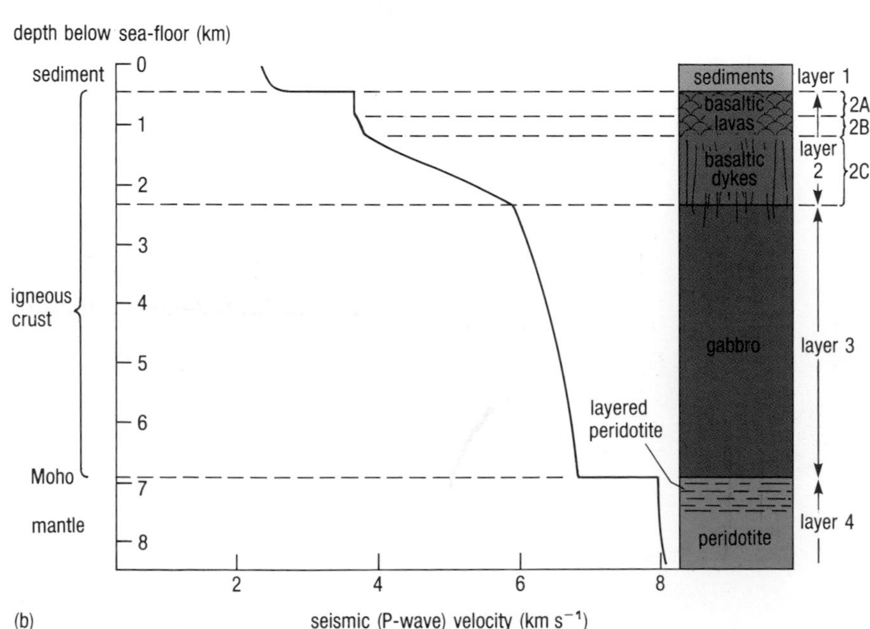

Figure 4.1 (a) Highly schematic sketch cross-section through the Atlantic Ocean with great vertical exaggeration. Velocities are given in km s^{-1}. See (b) for an explanation of the velocity layers.
(b) Typical seismic structure of the oceanic crust and upper mantle, showing how the velocity of propagation of P-waves increases with depth. The column is from the vertical line A–B in (a). P-waves are compressional waves, analogous to sound waves in air. The seismic velocity layers are traditionally matched with rock type as shown on the right. The layer 2 – layer 3 transition may sometimes occur at a shallower depth than suggested here, within the part of the crust made up of basaltic dykes (sheets of rock that were intruded vertically); see text for discussion. The individual rock types named in the column are illustrated in Figure 4.2.

The following points summarize the layered structure of the ocean floor:

(i) The igneous layers (2–4) are formed by processes at spreading axes, and are progressively buried by increasing thicknesses of sediments constituting layer 1, which eventually become consolidated into sedimentary rock.

(ii) Layers 2 and 3 constitute the igneous oceanic crust. The normal total thickness of these layers is 6–7 km. It may be less near transform faults and fracture zones, or more near hot spots such as Iceland.

(iii) The two most important changes in seismic velocity are at the base of the layer 1 sediments and at the base of layer 3. The latter is the Mohorovičić discontinuity (the **Moho**), where gabbro at the base of the crust rests on peridotite at the top of the mantle. Note that this occurs well within the lithosphere, so it is *not* the base of the tectonic plate (Figure 2.1).

(iv) Within layers 2 and 3 there are no definite seismic discontinuities, but rather a general increase of seismic velocity with depth. Subdivisions within layer 2 and the distinction between layers 2 and 3 are defined by changes of velocity *gradient* (changes in the rate of increase of seismic velocity with depth), and not by abrupt changes of seismic velocity.

(v) Despite their different appearances (Figure 4.2), newly formed rocks of layers 2 and 3 are very similar in their overall mineralogical (and chemical) composition; in order of decreasing abundance, the constituent minerals are plagioclase feldspar, pyroxene and olivine. However, the crystals forming the gabbro are larger than those in the dykes, which in turn are larger than in the lavas (this is a result of different cooling rates; slow cooling yields large crystals, whereas rapid cooling permits only small crystals to develop). The top of layer 4 accounts for the seismic discontinuity of the base of layer 3 (the Moho). The principal minerals of the peridotite of layer 4 are olivine and pyroxene.

(vi) The subdivisions of layer 2 shown in Figure 4.1 cannot always be recognized, but where they are, drilled samples suggest the following explanation. In sub-layer 2A, the lavas have many fractures and voids. In sub-layer 2B, the lavas are generally less broken and the voids are mostly filled with clays and other alteration minerals. This is partly because as soon as oceanic crust is formed it becomes subject to chemical alteration and metamorphism, by interaction with seawater (as we shall see in Chapter 5). Thus, it seems likely that the seismic velocity distinctions between layers 2B and 2C, and between layers 2C and 3, are often due to differences in the intensity of alteration and metamorphism, and/or progressive closure of cracks and pore spaces at greater depths and therefore greater pressure, rather than to differences in rock type. For example, ODP Hole 504B (in the Pacific Ocean, near the northern edge of the Nazca Plate, 200 km south of the spreading axis that separates it from the Cocos Plate, Figure 2.2) terminated 1.8 km into the igneous crust, apparently right at the base of the dyke layer, but already 0.6 km *below* the layer 2–layer 3 transition, which in this case was found to correspond to an increase in metamorphic grade (i.e. the degree of metamorphism) with depth.

(vii) Seismic velocities increase more rapidly with depth in layer 2 than in layer 3. Seismic velocities in layer 2 also tend to increase with age, i.e. with distance from the ridge, as the remaining cracks and pore spaces become infilled with minerals formed as a result of the interaction between rock and seawater.

QUESTION 4.1 Why should fractured lavas have lower seismic velocities than equivalent unfractured rocks?

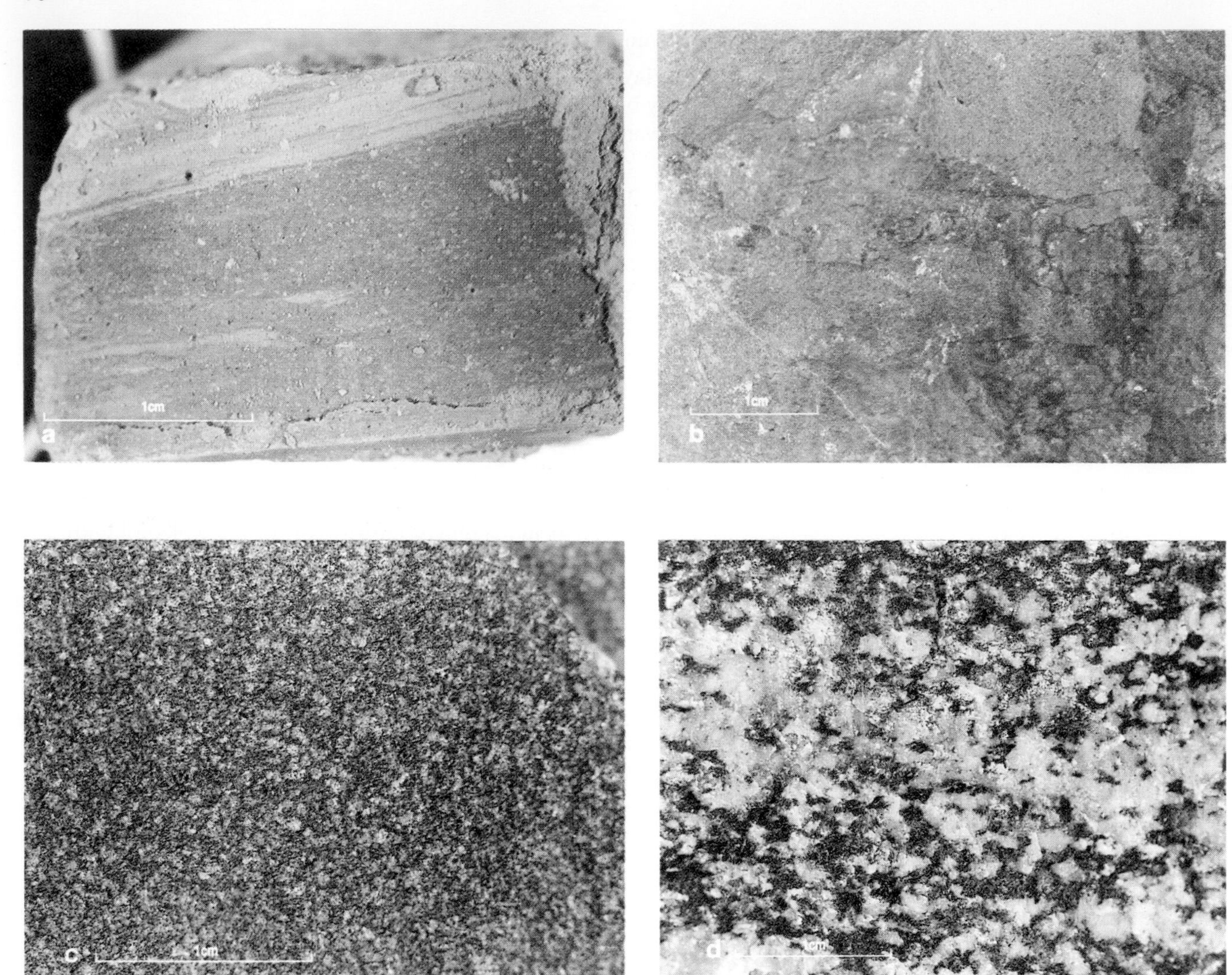

Figure 4.2 Colour photographs of sediments and igneous rocks of the oceanic crust and mantle.

(a) Deep-sea muds (layer 1).

(b) Fragment of basaltic lava from layer 2 (crystals too small to see at this magnification, because of rapid cooling after eruption onto the sea-floor).

(c) Fragment of basaltic dyke from layer 2 (discernible crystals, a fraction of a mm in size, indicative of less rapid cooling than the lava).

(d) Gabbro from layer 3 (large crystals, several mm across, as a result of slow cooling and crystallization at depth).

(e) Peridotite from layer 4.

4.1 THE FORMATION OF OCEANIC LITHOSPHERE

The axes of the ocean ridge systems are the most active volcanic zones on Earth, where spreading axes generate new oceanic lithosphere at rates of between 10 and 200 km per million years (1–20 cm yr^{-1}). The total volume of oceanic crust (layer 2 plus layer 3) produced at spreading axes has averaged in the region of 16–26 km^3 per year throughout the past 150 Ma.

Igneous processes at ridge crests are remarkable in other important respects. Unlike volcanoes elsewhere, which are active for relatively short periods of geological time, the volcanic activity at ocean ridges is episodically continuous for much of the lifetime of an ocean basin, which may be as long as a few hundred Ma. Moreover, the generation and eruption of magma occurs usually in such a consistent fashion that layering can be traced over hundreds of millions of square kilometres of ocean floor. Although much still remains uncertain about the sea-floor spreading process, it has become apparent that it is the diverging motion of plates at spreading axes that causes the igneous activity, rather than the igneous activity that forces the plates to move apart.

Despite uncertainties over the details, there is broad agreement on the overall mechanism (Figure 4.3). Beneath an oceanic spreading axis, asthenospheric mantle is drawn upwards so as to prevent a gap opening between the diverging lithospheric plates. As the asthenosphere rises past a depth of a few tens of kilometres, the steadily decreasing pressure triggers up to 10% partial melting. Although the material that is being partially melted has the composition of peridotite, the resulting melt (magma) has the composition of basalt. This basaltic magma is less dense and more mobile than the unmelted residuum, so it percolates upwards and eventually solidifies to form the crust. The unmelted residuum of the mantle (which is still peridotite, though depleted in those elements that are concentrated in basalt) accretes to the lower part of the diverging plates, to form the rest of the oceanic lithosphere.

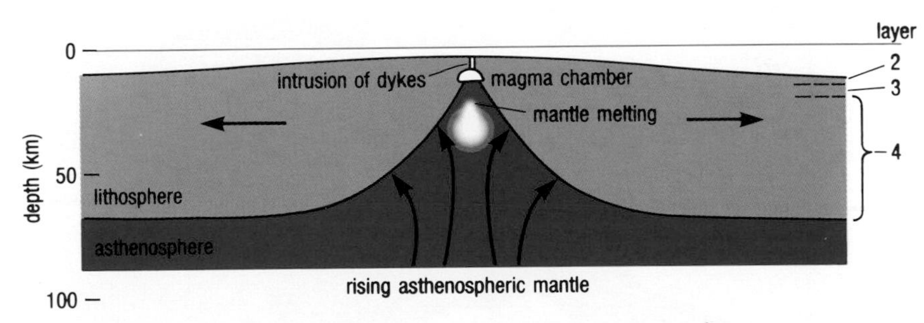

Figure 4.3 A cross-section showing formation of the lithosphere at a spreading axis. Rising asthenosphere fills the gap between separating plates. Some of it melts, ultimately giving rise to lava eruptions at the sea-floor and the formation of new oceanic crust. The remainder accretes to the edges of the lithosphere. A large magma chamber is rarely present; very often the gabbro layer forms from a crystal-rich mush (see text).

Exactly how the crust is generated beneath the ridge crest has been a matter of controversy for decades. The comparatively large size of the crystals in the gabbro layer (Figure 4.2(d)) shows that they are a product of slow cooling. It was once thought that in most places the magma collected in large, shallow **magma chambers**, several km in size, before crystallizing. Accumulation of crystals on the floor and walls of these chambers would give rise to the solid gabbro part of the crust. However, it is now believed that, often, the magma has begun to crystallize during its ascent from the source region, and arrives in the crust as a mixture of crystals and liquid. More importantly, it typically arrives in small batches, which mix with the (by then) even more crystallized magma that arrived in earlier batches. Thus, in most places, in the axial zone the gabbroic layer is a **crystal mush**, consisting of crystals mixed with only about 10% of magma, and lacks any large bodies of liquid magma. Old crystal mush cools and becomes entirely solid as sea-floor spreading carries it away from the axial zone, while the spreading axis is continually replenished by fresh batches produced by partial melting below.

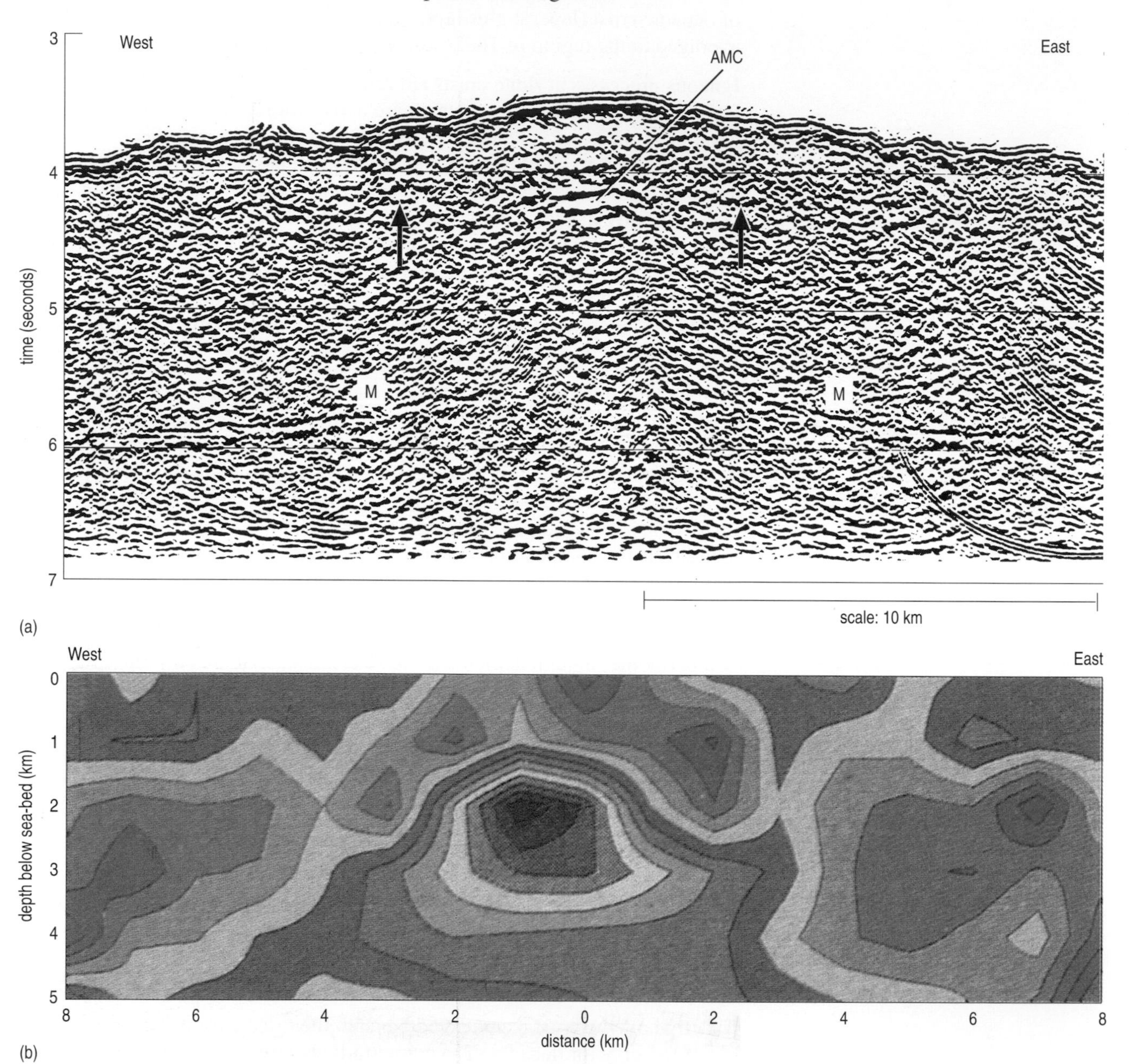

Figure 4.4 (a) Seismic reflection cross-section across the East Pacific Rise at 9° 30′ N. AMC marks the reflection off the roof of an 'axial magma chamber', which is probably a lens-like body at the top of the gabbro layer, rather than a thick magma chamber. The arrows on either side mark weaker reflections, perhaps representing the frozen (fully crystallized) top of this body, further from the axis. M marks the reflection from the Moho, which is not detectable directly below the axis because of the intervening AMC feature. (As is usual in seismic reflection, the vertical axis is labelled in seconds between transmission of the signal and receipt of the echo – known as two-way travel time.)
(b) Cross-section showing P-wave velocity structure across the East Pacific Rise at the same site as (a), derived by seismic tomography (see p.73). Pale blue represents normal P-wave velocity at each depth, and green, yellow and red tones represent progressively greater retardation of seismic velocity. In this case the maximum retardation is about 1.5 km s^{-1}. P-waves would be slowed down by about 3 km s^{-1} if travelling through a completely molten magma, and by proportionally less when travelling through a mixture of crystals and liquid. Note that the section in (b) corresponds to the upper portion of (a) and that the horizontal scales are different.

This model has been arrived at as a result of searching for reflections of seismic waves from the roof of any body of magma (Figure 4.4a), and from studies of the propagation of seismic waves through the crust at spreading axes, where the amount and location of magma can be inferred by the extent to which seismic velocities decrease (a technique referred to as seismic tomography, Figure 4.4b). Such studies indicate that typically below fast-spreading ridges there are lenses of magma a few hundreds of metres thick overlying a crystal mush zone, whereas slow-spreading axes have a narrower zone of mush and ordinarily lack persistent magma bodies of any kind. This is not to say that large magma chambers never occur, but they are probably ephemeral features.

A notable exception to the magma-free view of slow-spreading ridges is provided by the Mid-Atlantic Ridge at 57° N, where a magma lens 400 m thick, 4 km wide and at least 15 km long has been discovered by means of electromagnetic sensors deployed nearby on the sea-floor. Some scientists have suggested that more extensive deployment of sea-floor-mounted sensors would detect magma bodies elsewhere on the Mid-Atlantic Ridge too, and that the reason why they are not revealed by seismic reflection or seismic tomography surveys is that the signal is scattered because of the rough topography typical of this slow-spreading ridge. However, for the present it is safer to assume that bodies of magma (as opposed to crystal mush) are the exception rather than the rule on the Mid-Atlantic Ridge. If we accept this, then the cross-sections in Figure 4.5 show the typical structures of fast- and slow-spreading axes.

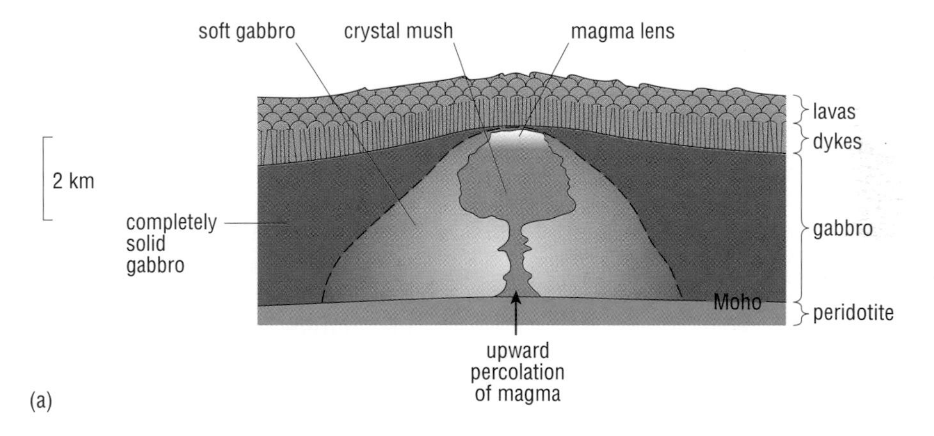

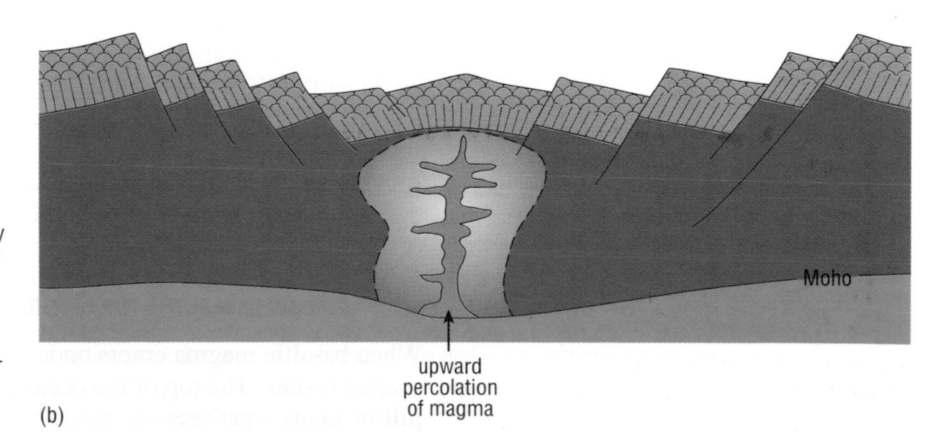

Figure 4.5 Models for the generation of crust at (a) fast- and (b) slow-spreading axes, based largely on seismic data, and each drawn to the same scale and with the same colour coding. Only the fast-spreading axis has an identifiable body of magma, and this has been drawn in a manner consistent with the evidence in Figure 4.4. In the case of the slow-spreading axis, dykes that were initially vertical are rotated by faulting as sea-floor spreading progresses. *Note:* These diagrams are at a very different scale from that of Figure 4.3.

(a)

(b)

Figure 4.8 (a) Pillow lavas in the Oman, or Semail, ophiolite of Oman, Arabia. Although many of the pillows here are seen in cross-section, the similarity to those shown in (b) is unmistakable.
(b) Pillow lavas photographed from a submersible on the Mid-Atlantic Ridge near 37° N; the field of view is about 3 m across.

Observations from submersibles, deep-tow cameras and cores recovered from boreholes show that the volcanic layer does not consist entirely of pillow lavas. Generally speaking, the faster a lava flow is erupted, the less effective is the surface chilling effect of the water and so the less pillow-like the resulting flow surface, which takes on a lobate and thin sheet-like morphology as eruption rate increases. In addition, pillows can easily fragment, so there are often local accumulations of volcanic debris (breccia) made of broken pillows and fragments of chilled glassy rind. As you will see in Section 4.1.3, non-pillowed sheet flows are particularly common at fast-spreading axes.

The formation of lavas thus makes for an irregular sea-floor, and tectonic and other processes combine to make the topography even more rugged (cf. Figures 2.11 and 2.12).

4.1.2 WHY A MEDIAN VALLEY?

Many people find the presence of a median valley confusing, because in cross-sections (Figures 2.12 and Figure 4.5b) it looks as if the floor of the valley has dropped *down* by movements across the faults that bound it. However, the axis of the valley is where rock is being added to the sea-floor, so the valley floor has never had the chance to be higher than we see it. This means that the faults bounding the median valley must actually mark where sea-floor rises *upwards* to its isostatically controlled equilibrium height, having been displaced away from the axis as a result of injections of new increments of crust at the axis. The simplest way to understand this is to appreciate that the age–depth relationship (whereby younger crust is hotter and therefore more buoyant than older crust, Section 2.3.2), applies only where the full thickness of the crust has become rigid. In the axial zone of a slow-spreading ridge, the carapace of lavas and dykes is not adequately supported by the weak and mushy gabbros beneath it, which is why the floor of the median valley is lower than the adjacent parts of the ridge structure.

According to this logic, fast-spreading ridges, which are underlain by a wider weak and mushy zone of gabbro, ought to have median valleys at least as wide as those along slow-spreading ridges. In fact, they have poorly developed or non-existent median valleys (Question 2.5, Figure 4.5a). The explanation for the *lack* of a large median valley on a fast-spreading ridge may be that fast spreading is usually associated with a particularly prolific magma supply, which maintains a sufficient volume of gabbro mush below the axis to keep the ridge inflated. The complex interplay between the rates of spreading and magma supply is something to which we shall return in Section 4.2. An alternative explanation (which need not concern us further) involves hotter mantle below fast-spreading ridges, which reaches isostatic equilibrium more quickly (closer to the ridge) than the relatively cooler mantle below a slow-spreading ridge.

4.1.3 FORMATION OF THE VOLCANIC LAYER: TWO CASE STUDIES

With the exception of isolated occurrences of deeper rock types revealed in the walls of faults, the volcanic layer of the oceanic crust is the only one that can be studied directly. This is because it is exposed over most areas of the ocean floor that are young enough not to have become buried by sediments. In this Section, we will look at the nature of the volcanic layer as seen in two contrasting areas: on the slow-spreading Mid-Atlantic Ridge and on the fast-spreading East Pacific Rise.

The Mid-Atlantic Ridge from 20° N to 40° N

The Mid-Atlantic Ridge has been most thoroughly investigated in the stretch extending south-westwards from near the Azores, between about 40° N and 20° N. This is not surprising, because the waters here are reasonably close to research bases in North America and Europe and are more hospitable than those further north in the Atlantic. Detailed study was begun in 1971 with Project FAMOUS (*F*rench-*A*merican *M*id-*O*cean *U*ndersea *S*tudy), which focused on a 10 km length of the spreading axis some 640 km south-west of the Azores. The ridge is atypically shallow here, with the floor of the median valley being at a depth of only about 2600 m. This was an important consideration in choosing the study area, because several manned submersibles were employed (including *Alvin*, Figure 1.22) that would have been unable to operate at more typical ridge depths. The photographs in Figures 4.7(d) and 4.8(b) were in fact taken through the window of *Alvin* in the FAMOUS area, and these and other photographs confirmed the widespread occurrence of pillow lavas and showed that they have usually been erupted from discrete vents, building up to form abyssal hills, several hundred metres across.

Project FAMOUS also used narrow-beam echo-sounding (Section 1.1.2) and *GLORIA* side-scan sonar (Section 1.2), but this was before the days of swath bathymetry and deep-tow devices. These and other technological advances made it possible for subsequent projects to work on the deeper, and therefore more typical, portions of the ridge further to the south-west. One such region is that bounded by the Kane and Atlantis transform faults, shown in Figure 4.9.

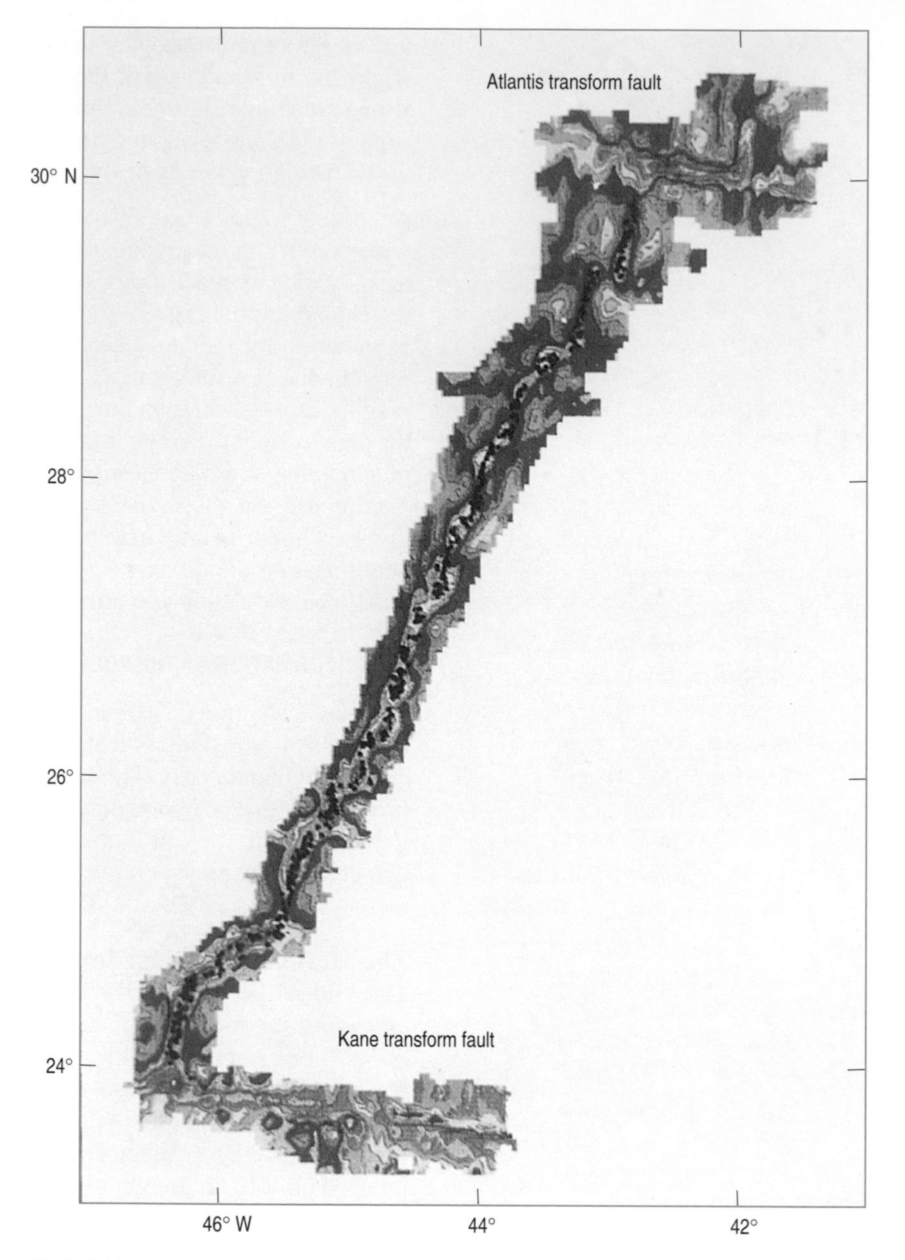

Figure 4.9 Map compiled using *Sea Beam* swath bathymetry, showing depths on the Mid-Atlantic Ridge between the Kane and Atlantis transform faults. Contour interval is 250 m. Depths less than 2500 m are in yellow, 2500–3250 m in red, 3250–4500 m in green, and deeper than 4500 m in blue. Black dots mark all the abyssal hills on the median valley floor more than 50 m high (a *TOBI* side-scan sonar image of one of these abyssal hills at 28° 50′ N is shown in Figure 2.18). Abyssal hills are just as closely spaced on the steep walls of the median valley, but these are disrupted by faulting and are not indicated on this map. Note that the area surveyed is confined to the vicinity of the median valley and the upfaulted regions on either side (compare profile 2 in Figure 2.12, which is similar). To either side of the area surveyed, the sea-floor gets gradually deeper, so although this map shows a valley, it must be borne in mind that this is the median valley at the crest of the Mid-Atlantic Ridge.

QUESTION 4.3 Locate the linear deep feature trending approximately east–west at about 23° 50′ N on Figure 4.9. Is this part of the median valley of the Mid-Atlantic Ridge? If not, what is it?

Concentrating on the median valley, it is apparent from the abundance of abyssal hills in Figure 4.9 that in this region at least, much of the volcanic part of the oceanic crust is constructed by the formation of abyssal hills at a succession of volcanic vents along the spreading axis. Figure 4.10 shows a detailed *Sea Beam* bathymetric map from within the area of Figure 4.9, and an interpretation of it. This shows younger volcanic abyssal hills overlapping and partly burying older ones, and demonstrates that the active volcanism is concentrated along the middle of the median valley. As a result, the volcanic products form what is known as an axial volcanic ridge. It also shows that abyssal hills are not the only volcanic features; at least half the sea-floor in this area is covered by lava flows that have erupted from fissures running parallel to the spreading axis, to form low hummocky ridges. Most abyssal hills and hummocky ridges mapped on either side of the axial volcanic ridge were formed by eruption of lava that occurred *on* the axial volcanic ridge, but have now been displaced to one side or the other (or, more rarely, split in two) by sea-floor spreading. However, some volcanism, for example that which formed the abyssal hill nearest the west wall and the associated hummocky ridge, may occur a few kilometres off-axis.

How can this pattern of volcanism be explained? We have already seen (Section 4.1) that seismic data make untenable the old idea of large magma

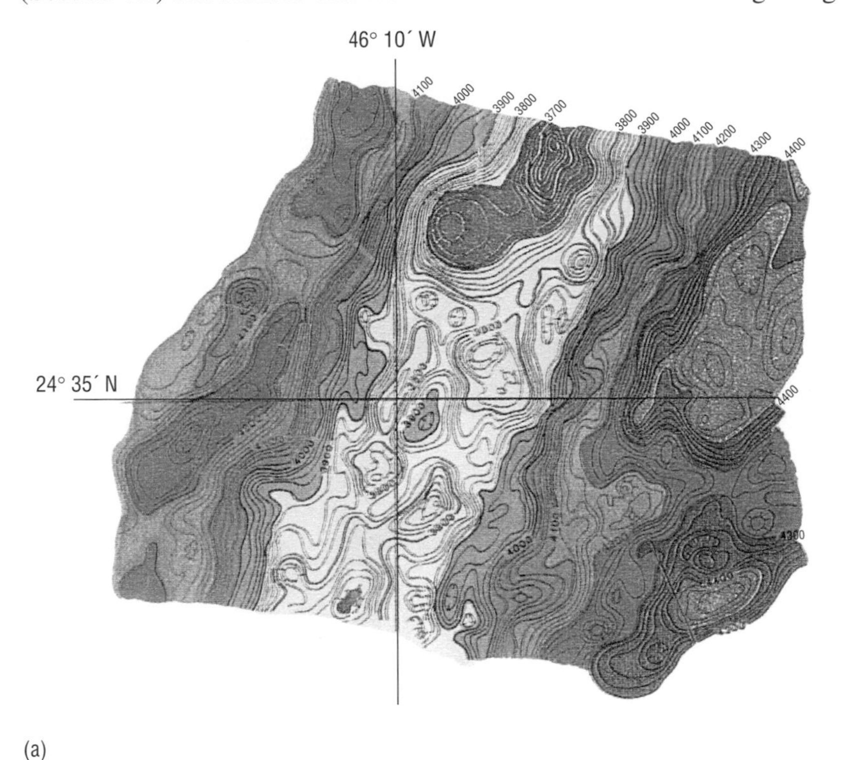

(a)

(b)

Figure 4.10 (a) Detailed *Sea Beam* bathymetric map of a portion of the median valley floor of the Mid-Atlantic Ridge, plotted at 20 m contour intervals. The axial volcanic ridge is the shallowest feature on the valley floor, and is shown in yellow and red. Tick marks on closed contours point downhill.
(b) Interpretation of the area shown in (a). Volcanic abyssal hills are shown with slopes hatched and shaded brown. The irregular elongated bands represent lavas erupted from fissures running parallel with the ridge; the younger the lavas the paler the tone. These fissure-fed lava flow fields create low ridges, and the crests of the youngest ones are marked by lines crossed by tick marks. The east and west walls (labelled) mark the edges of the median valley, where the sea-floor begins to be faulted upwards.

chambers typically underlying spreading axes, especially at slow-spreading axes like the Mid-Atlantic Ridge. However, a magma supply is clearly required to account for the lavas. Figure 4.11 shows a possible arrangement in which each volcanic abyssal hill or hummocky ridge is fed from a separate local source of magma.

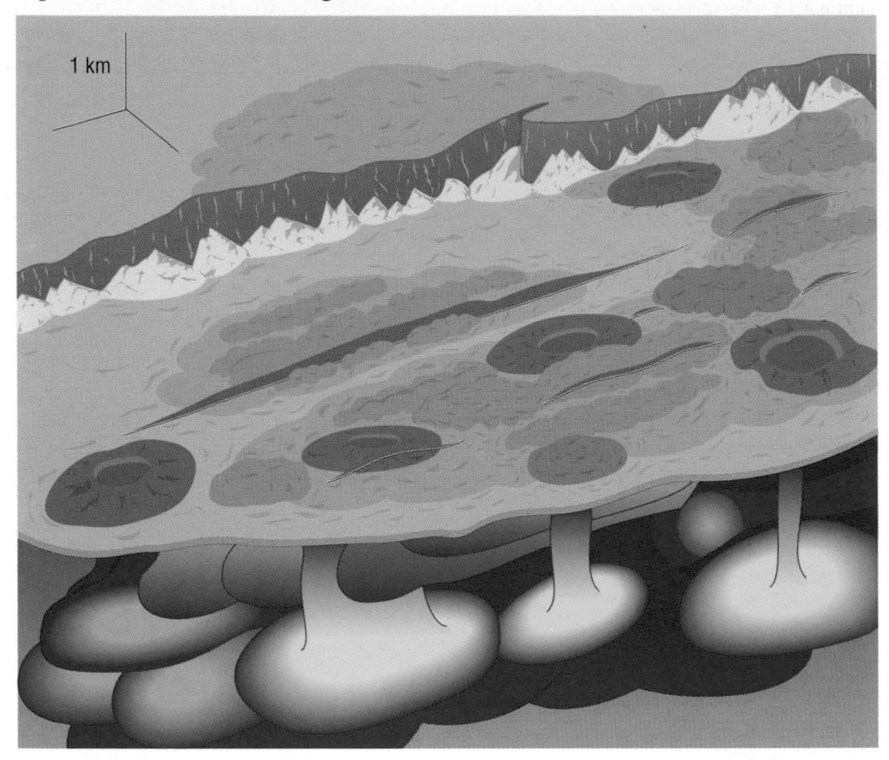

Figure 4.11 Schematic cut-away perspective view covering just over half the width of the median valley of the Mid-Atlantic Ridge in an area similar that in Figure 4.10. Volcanic abyssal hills are shown in dark brown, and lavas erupted from fissures are shown in pale green to light brown (youngest = brownest). The wall at the back represents the fault scarp bounding the far side of the median valley; some debris fans are shown along it. A few faults are shown on the valley floor, cutting through the volcanically generated topography (these are revealed by deep-tow side-scan sonar images). The lower part of the Figure is an interpretation of the situation below the sea-floor; vertical red sheets (dykes) connect to the fissures from where the lavas forming the hummocky volcanic ridges are erupted, and more pipe-like feeders (but still elongated parallel to the axis) underlie the volcanic abyssal hills. The dykes and pipes are supplied from small flat 'magma bodies' (each one active at a different time).

QUESTION 4.4 In view of the evidence discussed in Section 4.1, would you expect the small flat magma bodies shown in Figure 4.11 to be entirely molten (i.e. just magma) or a mixture of magma and crystals?

Perhaps the most striking revelation of this sort of study is that active axial volcanism, and presumably the dykes which feed the lavas, occur in localized zones, and that there are intervals of local extension without volcanism after the extinction of one volcano (volcanic abyssal hill or hummocky volcanic ridge) and before the growth of its neighbour. As noted earlier, the frequency of dyke injection that you calculated in Question 4.2 is clearly an average valid only on time-scales of the order of 100 000 years and longer.

QUESTION 4.5 Given that the full spreading rate across the Mid-Atlantic Ridge near 25° N is 25 mm yr^{-1}, and taking the width of the median valley between the 'west wall' and 'east wall' in Figure 4.10(b) to be representative of the average spacing between the major faults bounding the median valley, on average how long is the interval from the eruption of lavas on the crest of the axial volcanic ridge until they begin to be faulted upwards?

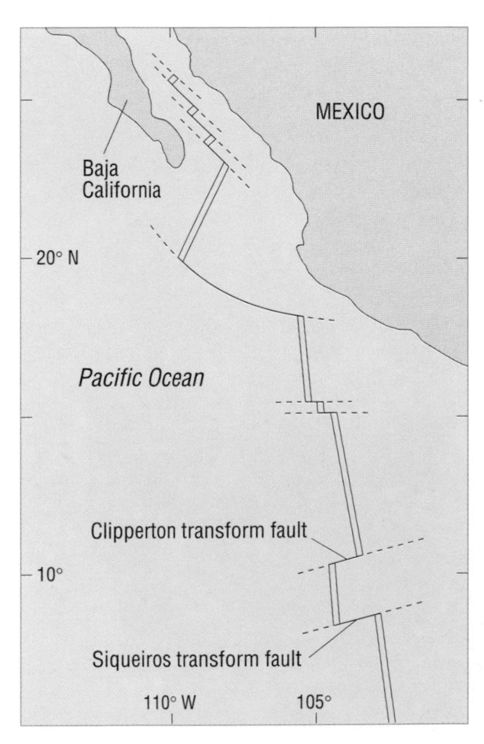

Figure 4.12 Map to show the location of the East Pacific Rise near 9°–10° N.

The East Pacific Rise from 9° N to 10° N

Not unnaturally, the East Pacific Rise has been most intensively studied where it lies closest to the North American coast. We will concentrate here on the stretch that lies between the Clipperton and Siqueiros transform faults (Figure 4.12). This portion of the East Pacific Rise is spreading at a half-rate as fast as 55 mm yr^{-1}, yet (like fast ridges in general) it seems to be abundantly supplied with magma. We have already seen seismic reflection and seismic tomography data that provide convincing evidence for a high-level axial magma chamber overlying a crystal mush zone (Figures 4.4 and 4.5(a)). In cross-section, the ridge takes the form of a broad rise (Figure 4.13). The median valley is so small that it can usually be mapped only with the aid of side-scan sonar images; *SeaMARC* was used in this area, and the images obtained show that, where present, the valley is only about 200 m wide and 20 m deep. This valley is thought to be formed by collapse following the withdrawal or eruption of magma, and is probably a temporary feature, likely to disappear next time the underlying axial magma chamber is re-inflated by a significant input of magma. To distinguish this sort of feature from the much larger and more persistent median valleys on slow-spreading ridges (and by analogy with volcanoes on land) it is sometimes referred to as the 'axial summit caldera'.

This part of the East Pacific Rise makes a particularly good case study because it includes one of the few examples of an area where the ocean floor has been documented in detail both before and after a volcanic event

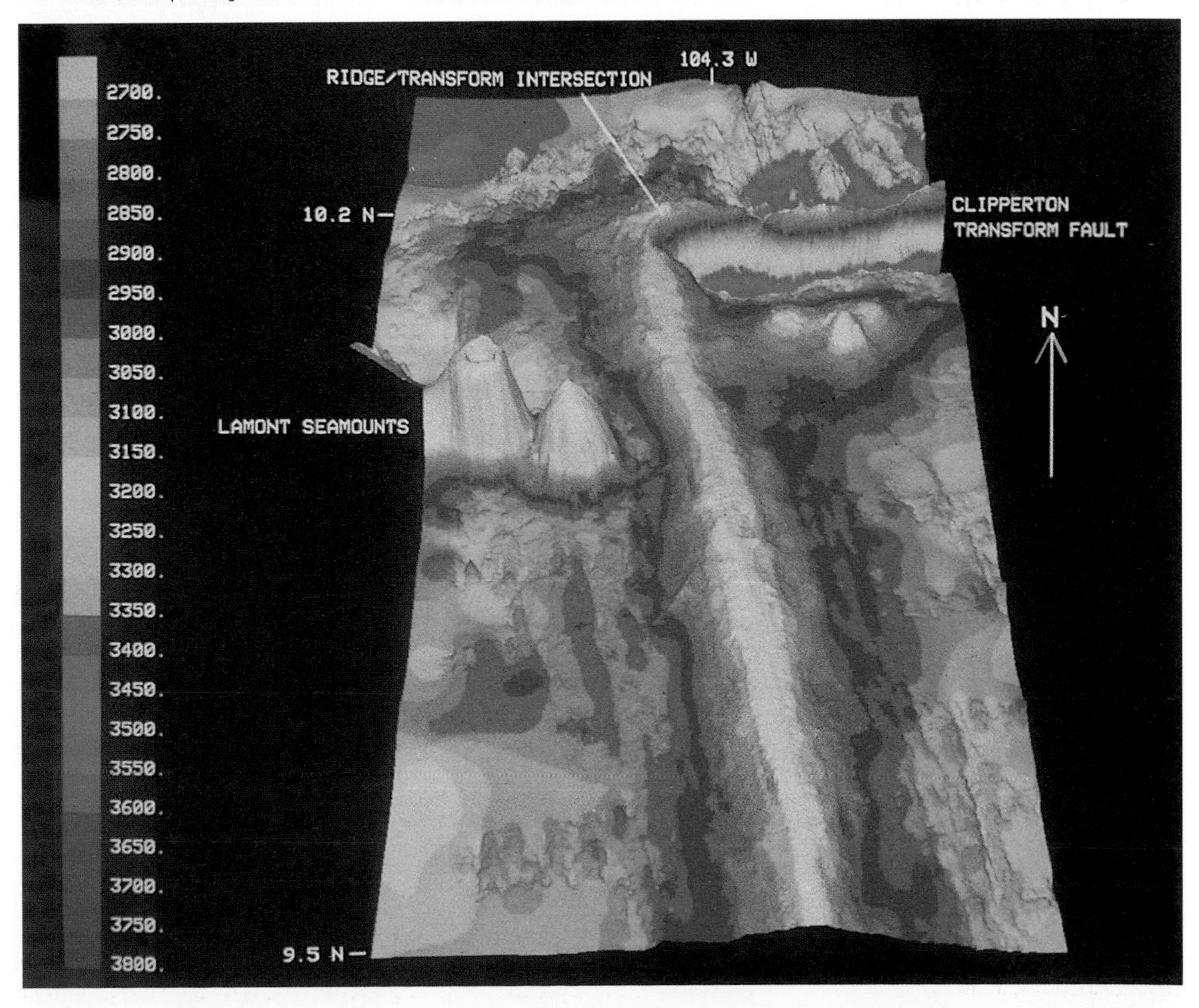

Figure 4.13 Perspective view of the East Pacific Rise between 9° 30′ N and the Clipperton transform fault, compiled from *Sea Beam* swath bathymetry data. From north to south, the distance covered is about 100 km. Depths are colour coded, from blue (deeper than 3500 m) through green, yellow and red to pink (shallower than 2700 m). The ridge crest has a narrow median valley along about half its length, but it is too small to show at this scale. Note the off-axis bathymetric fabric trending parallel to the ridge, which is a result of faulting on the ridge flanks. The Lamont seamounts probably result from intermittent volcanism associated with a 'hot spot' (Section 2.5.3), and so are not indicative of the normal state of the spreading axis.

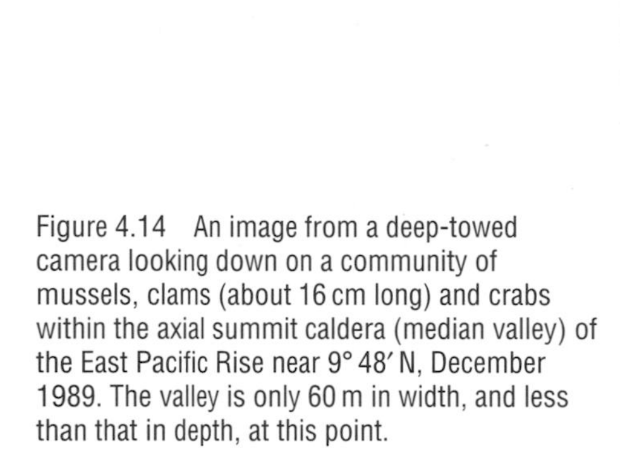

Figure 4.14 An image from a deep-towed camera looking down on a community of mussels, clams (about 16 cm long) and crabs within the axial summit caldera (median valley) of the East Pacific Rise near 9° 48′ N, December 1989. The valley is only 60 m in width, and less than that in depth, at this point.

associated with sea-floor spreading. In November and December 1989, an 83-km-long stretch of the ridge crest was mapped with deep-tow cameras. These showed many sites where warm mineral-rich fluids were emerging through fissures (hydrothermal vents) on the floor of the median valley. The vents were surrounded by mussels, clams, tubeworms and other invertebrates forming ecological communities supported by the chemical energy in the vent fluids (Figure 4.14). We will consider hydrothermal vents and the animals that live near them in more detail in Chapter 5, but for now the important point is that by the time a series of *Alvin* dives was begun in April 1991, most of the previously discovered communities on the East Pacific Rise at 9°–10° N had been overrun by newly erupted lava (Figure 4.15).

Figure 4.15 (a) Photograph from *Alvin* in April 1991, showing fresh slabby lava which has buried the community seen in Figure 4.14. White bacterial mats line the cracks in the lava surface, showing it has cooled down sufficiently for organisms to take hold. The field of view is about 5 m across.

(b) Photograph of what has become known as the 'Tubeworm Barbecue' site, taken from *Alvin* on 14 April 1991, near 9° 49′ N. Dead, toppled tubeworms (and other dead animals) litter the foreground, and fresh lava is visible in the background. The edge of the lava flow is no more than about 10 cm thick. Radiometric dating of samples taken from it suggest that it erupted between 26 March and 6 April 1991, which is consistent with the observation that there had not yet been time for scavengers to have re-entered the area to consume the remains. The field of view is about 5 m across.

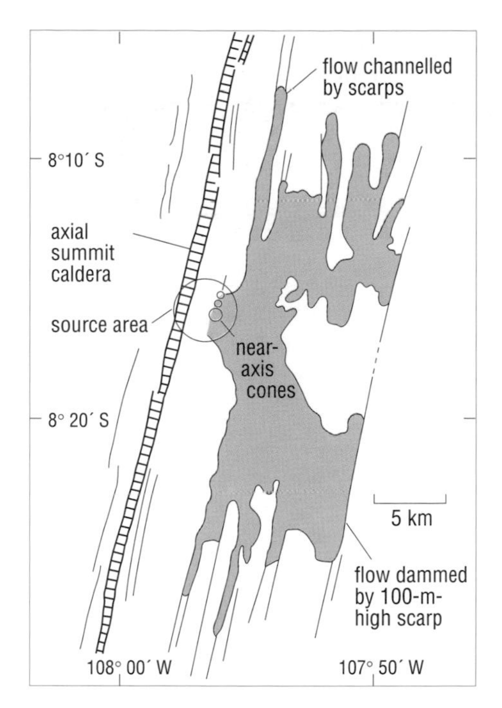

Figure 4.16 Map of a lava flow that was imaged by *SeaMARC* side-scan sonar in 1987, and estimated to be less than 25 years old at that time. Where it over-tops certain fault scarps, the flow must be ponded to a thickness of up to 70 m, but elsewhere it is probably 5–10 m in thickness.

The extent of the lava flows erupted in the 1991 event could not be determined by the *Alvin* dives. However, if they were similar to young flows imaged by side-scan sonar further south along the East Pacific Rise, then they may have escaped the confines of the axial summit caldera and spread out over one flank or the other to cover an area of the order of 100 km^2 (Figure 4.16).

QUESTION 4.6 On the basis of the descriptions in this Section, what are the main differences in styles of volcanic eruptions on the Mid-Atlantic Ridge and the East Pacific Rise?

We have now looked at differences in cross-sectional profile and in the style of lava eruption on fast- and slow-spreading axes. To complete our survey of the formation and structure of the oceanic lithosphere at spreading axes, we must now consider variations that occur *along* an axis.

4.2 SEGMENTATION OF OCEANIC SPREADING AXES

We have reviewed plenty of evidence to show that, at any time, volcanic activity is restricted to localized stretches of a spreading axis. Examples include the individual volcanic abyssal hills (about 2 km across) and hummocky volcanic ridges (only a few km long) on the Mid-Atlantic Ridge (Figures 4.10 and 4.11), and sheet flows of lava on the East Pacific Rise (extending for a few tens of km along axis, though their source regions are probably less extensive than this; Figure 4.16).

However, mapping the extent of individual erupted units actually tells us rather little about the situation deeper in the oceanic crust. Even if a spreading axis were underlain by a magma chamber or crystal mush zone 1000 km long, there would be no need for the dyke layer ever to rupture along this whole length at a single time, and to supply lavas along this whole length at once. Studies of the mineralogical and chemical composition of mid-ocean ridge basalts demonstrate that similar processes occur along the whole length of a spreading axis, but they neither confirm nor refute the possibility of a spatially continuous supply of magma. However, a variety of other evidence demonstrates that there are along-axis variations in the magma supply. These variations control the division of a spreading axis into **spreading segments**, each of which can be distinguished from its neighbour at either end.

There is one simple observation that makes it unlikely that a single magma chamber or crystal mush zone can run along the whole length of an ocean ridge axis. Can you suggest what it is?

It is simply the existence of transform faults. You have seen how transform faults divide ridges into segments. Figure 4.9 shows the Kane and Atlantis transform faults on the Mid-Atlantic Ridge (800 km apart), and Figure 4.12 shows the Clipperton and Siqueiros transform faults on the East Pacific Rise (220 km apart). Each of these faults offsets the ridge axis by at least 100 km. Transform faults with smaller offsets can be seen further north on Figure 4.12, and in fact transform faults with offsets as small as about 30 km are regarded as being representatives of the same kind of phenomenon, which is described as a **first-order discontinuity** in the spreading axis. Each length of axis between such transform faults can therefore be described as a **first-order segment**. Some, though certainly not all, of these transform faults were probably inherited from fractures along lines of weakness during continental break-up when the ocean began to open (Figure 2.15). However, whatever their origin, when we consider that axial crystal mush zones have been determined to be only a few kilometres wide on both slow- and fast-spreading axes (Figure 4.5), it is difficult to see how there can be any continuity in magma supply across offsets of such magnitude.

Further evidence for the axial discontinuity of magma supply comes from seismic reflection studies performed off-axis in the Atlantic. These suggest that the gabbro layer is thickest between fracture zones. It gets thinner, and may even disappear, near fracture zones. The only crust at many fracture zones appears to be dykes that propagated beyond the end of the crystal mush zone, and lavas fed by these dykes. This situation is illustrated in Figure 4.17. The implication is that magma supply is poor near transform faults, but more abundant near the centres of spreading segments.

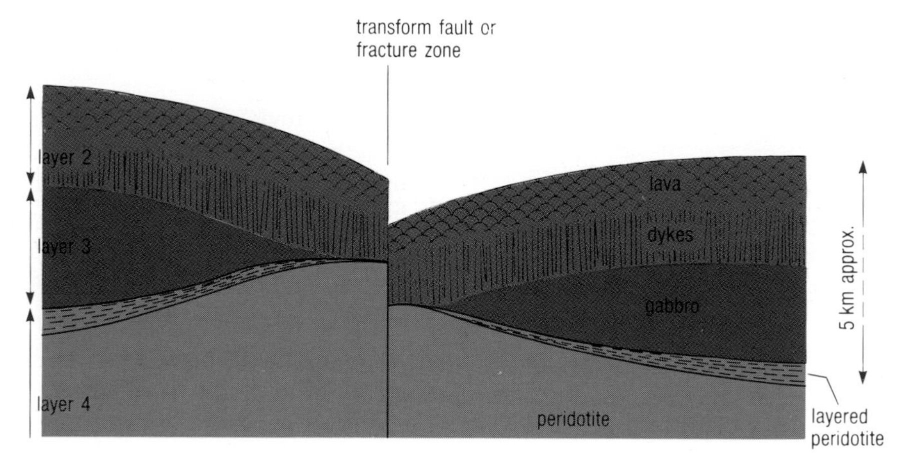

Figure 4.17 Schematic cross-section parallel to a spreading axis based on off-axis seismic reflection studies, showing how the gabbro layer appears to get much thinner, and may disappear entirely near a transform fault. The study area upon which this interpretation was based is in crust approximately 130 Ma old, offshore of North America and about 500 km south of the Kane fracture zone.

Thus, so far as first-order segmentation is concerned, zones of axial magma supply commonly do not extend right up to transform faults, and where they do the scale of the offset means that they are unlikely to be in contact with a similar zone beyond such faults.

More detailed studies reveal that spreading axes are frequently offset by smaller amounts too. Most of the smaller-scale features where this occurs are varieties of non-transform offsets, which were alluded to in Section 2.4. At this scale, a marked difference emerges between fast- and slow-spreading axes. We will look at fast-spreading axes first.

4.2.1 SECOND- AND THIRD-ORDER SEGMENTATION OF FAST-SPREADING AXES

Swath bathymetry surveys on the axis of the East Pacific Rise and other fast-spreading ridges reveal offsets of 2–15 km in places where there is clearly no transform fault. Instead, the axial crest begins to die away and a new axial crest begins to build up beside it. The places where axial crests overlap have been named **overlapping spreading centres*** or OSCs. Figure 4.18 shows a 1000-km length of the East Pacific Rise, where there are four transform faults and eight overlapping spreading centres, and Figure 4.19 shows details

* The term is derived from the use of 'spreading centre' to refer to an active stretch of spreading axis, which is not entirely appropriate because it implies spreading in all directions from a central point, whereas in fact spreading occurs outwards in opposite directions from a line.

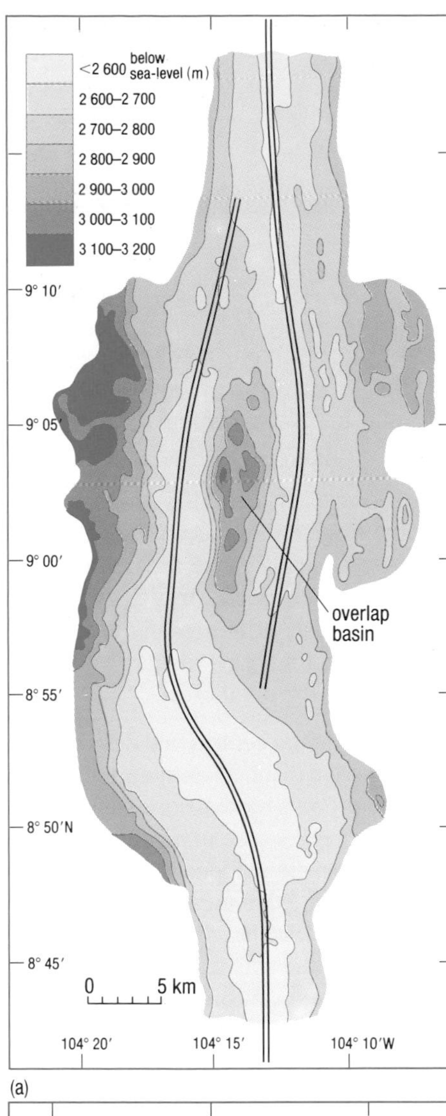

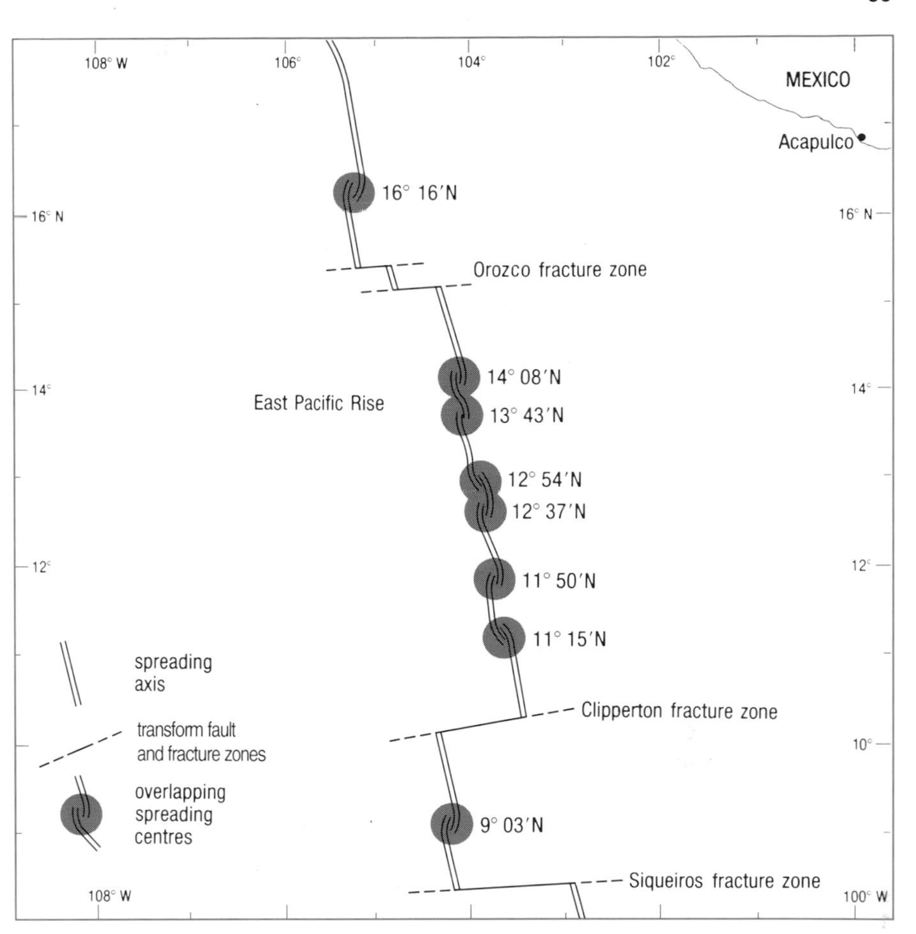

Figure 4.18 Overlapping spreading centres along a 1000-km length of the East Pacific Rise. The transform faults break the ridge into first-order segments, and the overlapping spreading centres further divide it into second-order segments.

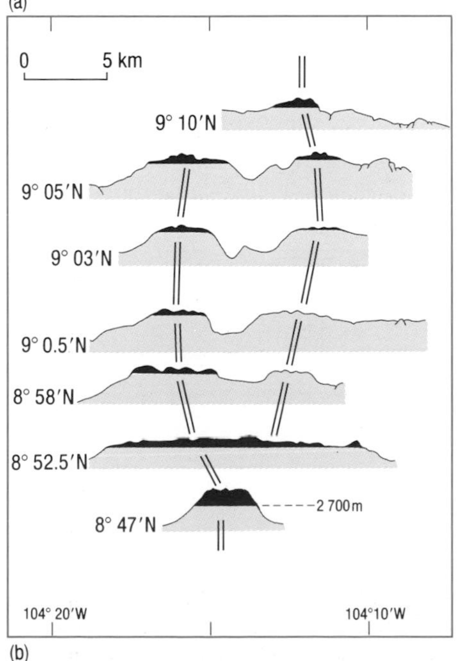

of the only overlapping spreading centre on the axis between the Clipperton and Siqueiros transform faults. Overlapping spreading centres are described as **second-order discontinuities**, and they sub-divide the axis into **second-order segments**.

Submersible observations show that the youngest lava flows usually occur on or very close to the axial crests, and it is clear that each crest (irrespective of whether or not it has a small median valley) must lie directly over a zone of dyke injection. The axial crests are shallowest near the middle of each second-order segment, suggesting that the rate of magma supply is greatest here. This is consistent with the observation that both volcanic and hydrothermal activity (see Chapter 5) are generally more intense near the middle of each second-order segment and die out towards the overlap zones. On the other hand, fracturing and small-scale faulting increase towards the overlap, presumably in response to stresses induced by the offset. Furthermore, where it has been mapped, seismic reflection from the roof of the supposed magma lens (Figure 4.4(a)) tends to die out close to the overlapping spreading centre.

Figure 4.19 (a) Bathymetric contours of an overlapping spreading centre on the East Pacific Rise. The axial crests are indicated by double lines.
(b) Topographic cross-sections through the same overlapping spreading centre. The shallowest areas (less than 2700 m depth) are shown in black. These tend to consist of the youngest lava flows. Vertical exaggeration ×4.

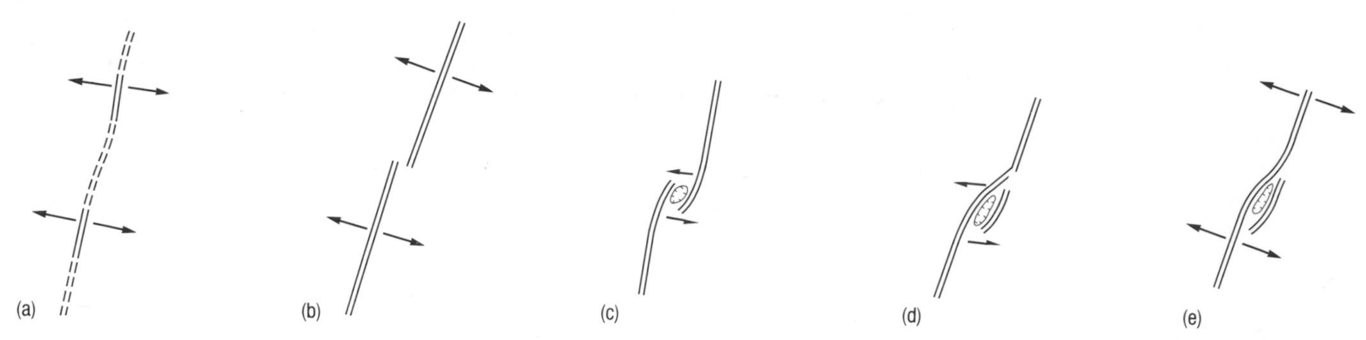

Figure 4.20 Possible model for the formation and evolution of an overlapping spreading centre. The double lines mark axial crests above zones of active dyke injection.
(a) Start of the cycle: ridge activity has been subdued for some time, due to lack of supply of magma.
(b) Fresh supplies of magma arrive in the crust, which causes the zones of dyke injection to propagate.
(c) Axial crests (and zones of dyke injection) overlap, and begin to curve towards one another as a result of the stress field.
(d) The more vigorous zone propagates and cuts off the overlapping portion of the other one. At this stage, the two axial crystal mush zones, possibly topped by a continuous magma lens (Figure 4.5), may have merged.
(e) The old overlap tip is abandoned, but remains as a topographic high. This is the end of a cycle. If magma supply wanes temporarily, the situation may return to (a). A complete cycle takes around 100 000 years.

Thus, it appears that the axial magma lenses or crystal mush zones must have breaks additional to those at transform faults. In fact, if the model shown in Figure 4.20 is correct, the gaps in the magma supply are responsible for the very existence of the overlapping spreading centres.

At a smaller scale still, about every 10 km along each second-order segment of the East Pacific Rise, the axial zone of the ridge either has a minor dog-leg with an offset measured in hundreds of metres or less, or else the trend of the axis is bent by a few degrees. These are **third-order discontinuities**, and mark **third-order segments**. Often these minor discontinuities in the axis are referred to as 'deviations from axial linearity', or, for short, 'devals'. More often than not, devals occur at local low points in the axial crest.

An important difference between overlapping spreading centres and devals is that the former produce bathymetric features that are distinctive enough to be recognized even after they have been displaced off-axis by sea-floor spreading. The abandoned overlap tip in Figure 4.20(e) would be a case in point. Often, a whole series of progressively older such features can be mapped in a belt leading away from an overlapping spreading centre, indicating that the cycle shown in the Figure has been repeated many times at more or less the same spot on the ridge. In contrast, devals can be recognized only on the axis itself, from which it is concluded that they are short-lived features, whose location and number vary on a timescale of less than about 10^5 years.

With devals, we appear to have arrived at the smallest scale at which magma supply is spatially discontinuous (segmented). Evidence for this is provided by seismic tomography, which suggests that the axial crystal mush zone does not extend across every deval (Figure 4.21), and by seismic reflection studies in which reflection from the roof of the supposed magma lens is lost near some (but not all) devals. Moreover, whenever sampling of mid-ocean ridge basalts along an axis reveals any sudden change in chemical or mineralogical

Figure 4.21 Cross-section *along* the East Pacific Rise, showing retardation of P-wave velocities. This is a seismic tomography interpretation at right angles to that in Figure 4.4(b) (which it crosses 1 km west of the zero datum), and is shown with the same colour code. These data suggest that the axial crystal mush zone does not extend across the deval at 9° 28′ N, though it may cross the 9° 35′ N deval.

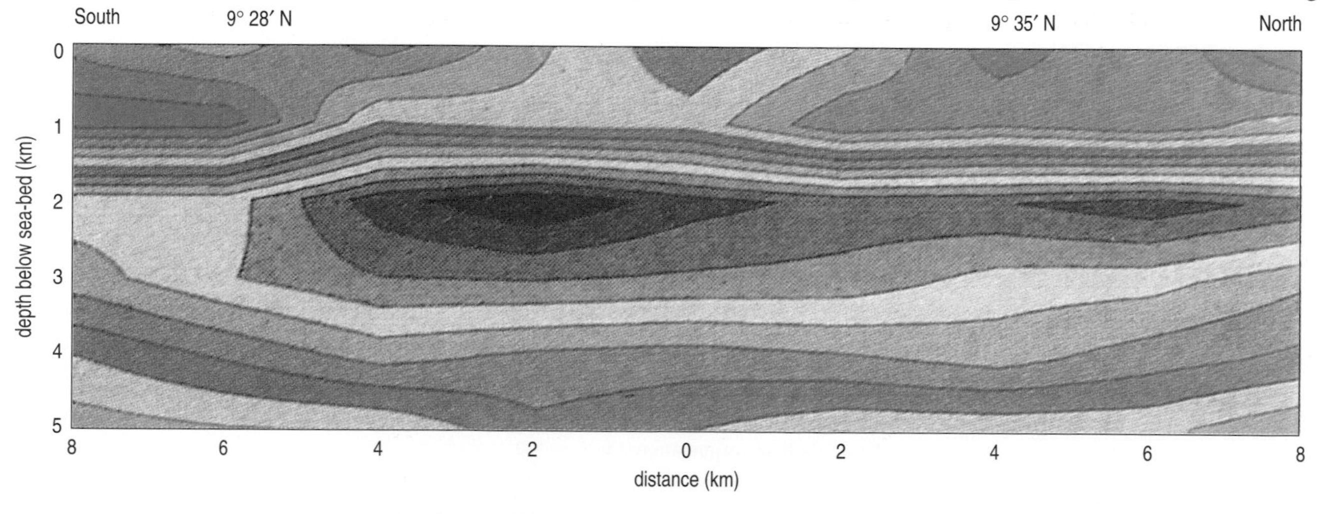

characteristics, it usually occurs at a deval or higher (i.e. first- or second-) order segment boundary. We will look at a possible explanation for magmatic segmentation shortly, but first we need to make a comparison with slow-spreading axes.

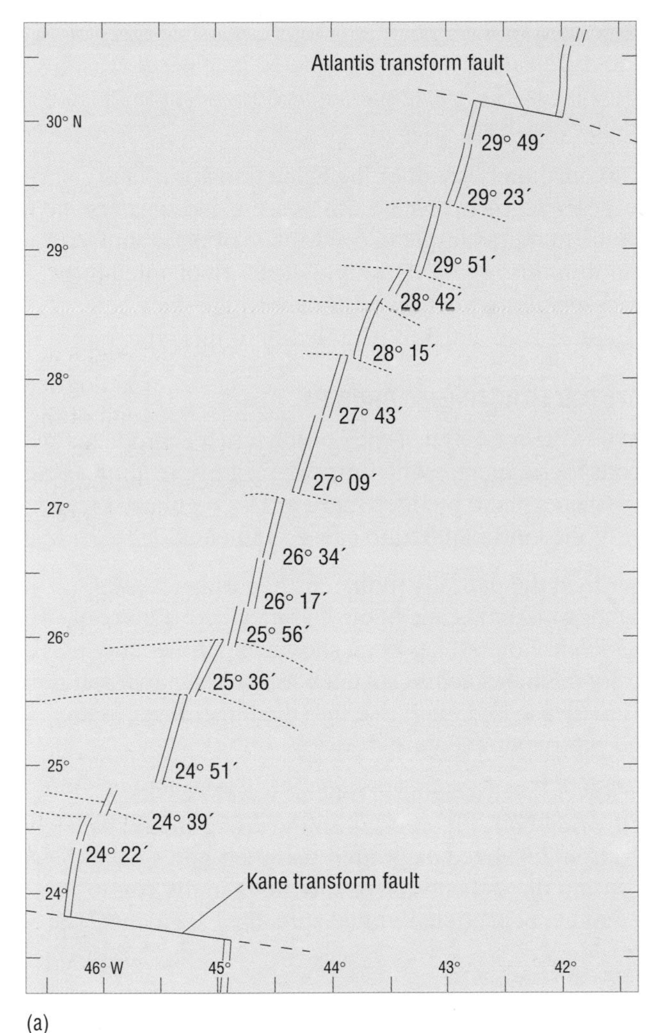

4.2.2 SECOND- AND THIRD-ORDER SEGMENTATION OF SLOW-SPREADING AXES

Overlapping spreading centres have not been identified on slow-spreading axes, so there are none along the Mid-Atlantic Ridge. Inspection of Figure 4.9, however, reveals a number of places between the Kane and Atlantis transform faults where the trend of the axis is disturbed. These discontinuities in axial trend are not small transform faults, and the edges of the diverging plates appear to behave non-rigidly here. The more obvious discontinuities are associated with bathymetric features that extend off-axis (Figure 4.22(a)), indicating a degree of longevity comparable with overlapping spreading centres, and like OSCs these are regarded as second-order discontinuities. Other discontinuities with no off-axis trace are directly comparable with Pacific-style devals and are therefore third-order offsets. Most of these non-transform offsets occur at places where the floor of the median valley is locally deep, depths being least near the middle of each segment (irrespective of whether they are second- or third-order, Figure 4.22(b)). This suggests greater magma supply near the middle of each segment. We have already noted a comparable axial depth relationship for segments on the East Pacific Rise bounded by OSCs and devals.

Figure 4.22 (a) Interpretation of the Mid-Atlantic Ridge in the Kane–Atlantis region, based mainly on *Sea Beam* bathymetric surveys (cf. Figure 4.9). The latitudes of second- and third-order discontinuities are indicated, to aid comparison with (b). Off-axis bathymetric traces are shown by dashed lines, and define the second-order discontinuities. (b) Along-axis depth variation in the Kane–Atlantis part of the Mid-Atlantic Ridge. Latitudes of discontinuities correspond with those shown in (a).

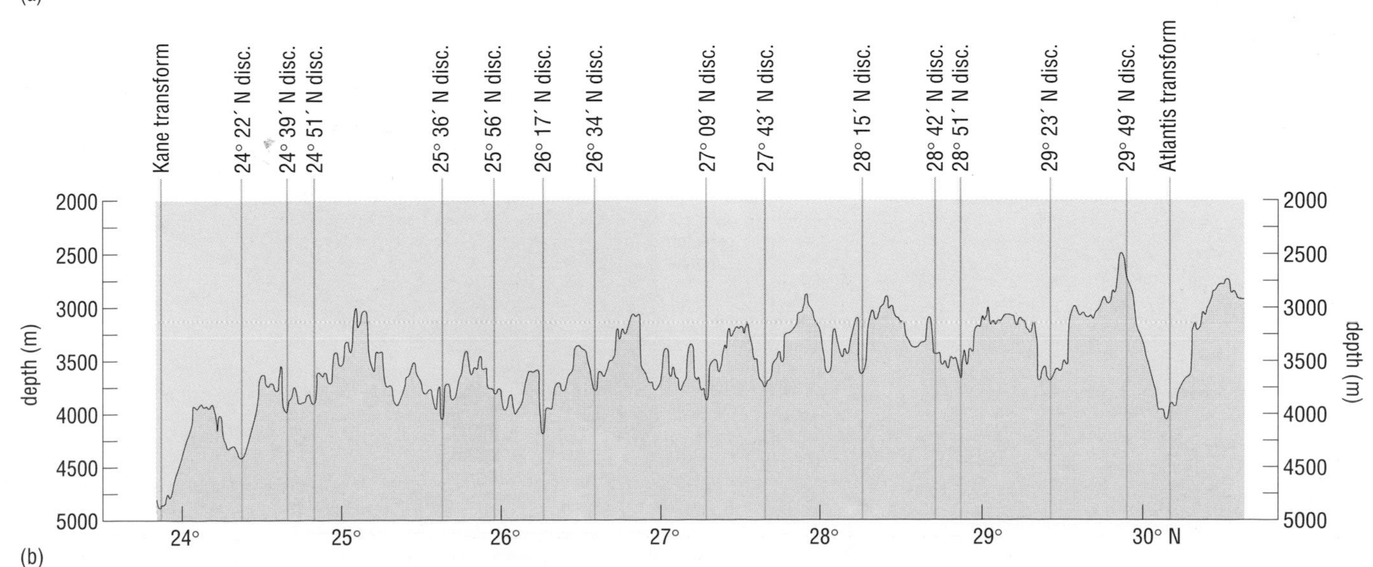

A particular characteristic of slow-spreading ridges is that it is not uncommon for sea-floor spreading in some second-order segments to proceed apparently by **amagmatic extension**. In these places, upwelling of the mantle asthenosphere below the zone of plate divergence appears to be so sluggish that little melt is produced, and much of it crystallizes within the mantle. The lithosphere created in these segments thus consists largely of peridotite, cut by basaltic dykes and overlain by isolated patches of crust in the form of basaltic pillow lavas. Such magmatism as does occur is concentrated in the middle of each segment.

The first-order segment immediately south of the Kane transform fault contains several second-order segments where this is seen, and samples retrieved by dredging confirm the widespread occurrence of peridotite exposed on the ocean floor in this region. The style of sea-floor spreading along much of the slower-spreading South-west Indian Ridge (between Africa and Antarctica, Figures 2.26 and 3.6) is probably of this type.

4.2.3 A PLAUSIBLE MODEL FOR LITHOSPHERIC GROWTH

It was stated in Section 4.1 (Figure 4.3) that the oceanic crust is produced by partial melting of asthenospheric mantle which rises beneath spreading axes. But how can the rising asthenosphere produce more or less regularly spaced zones of magma supply of the kind required to cause segmentation?

The lithosphere is denser than the partially molten asthenosphere; over geological time-scales, this situation is analogous to one where a layer of dense fluid overlies a less dense one. It has been shown experimentally that if the density and viscosity contrasts across such a boundary are appropriate, then the fluid from the less dense layer will rise into the upper layer in the form of regularly spaced protrusions (Figure 4.23(a)).

At a spreading axis, the lower layer is a linear zone of lower density asthenosphere beneath the ridge (cf. Figure 4.3), so the effect should be the same. Figure 4.23(b) is an idealized diagram illustrating how magma concentrations within the asthenosphere should naturally rise to form concentrations of magma beneath individual spreading segments. The

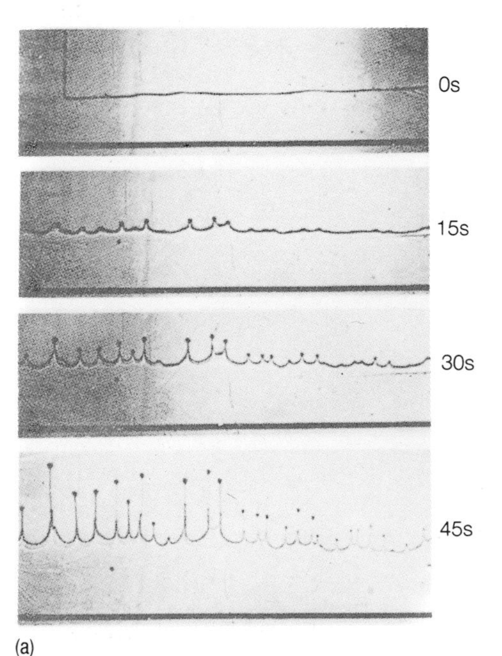

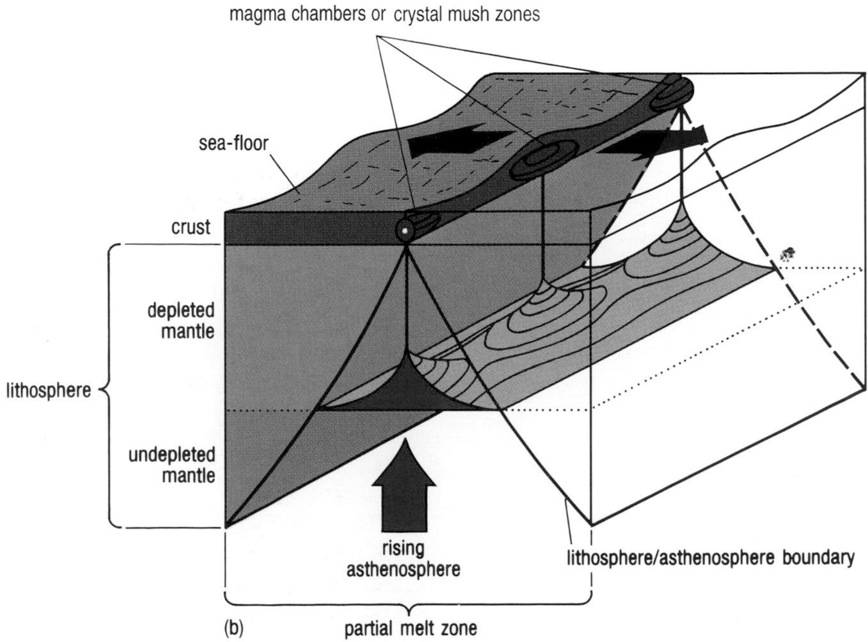

Figure 4.23 (a) Gravitational instability of a water–glycerine mixture (less dense) injected rapidly along a horizontal line into glycerine (more dense). A fairly regular spacing of rising 'blobs' of the less dense mixture is apparent after 45 seconds. The water–glycerine mixture represents partial mantle melts, and the individual blobs represent feeder zones for individual crystal mush zones (or magma chambers) in (b).

(b) Viscous asthenosphere rises beneath the boundary between two spreading lithospheric plates. Above a certain level (dotted lines), the rising asthenosphere passes into a zone in which partial melt can form, and will permeate the asthenospheric mantle at some level below the base of the lithosphere. Its lower viscosity and density cause the partial-melt zone to develop a gravitational instability leading to regularly spaced concentrations of melt. These percolate towards the surface to form discrete crystal mush zones (or magma chambers) in the crust, giving rise to segmentation along spreading axes. The asthenosphere continues to rise viscously, and on cooling becomes rigid and joins the lithosphere which thickens away from the spreading axis. The bold lines mark the boundary between lithosphere and asthenosphere.

spaces between the segments will be starved of magma, topographically lower, and have thinner crust.

Once a magma chamber or crystal mush zone has been formed in this way, it is likely to continue to be fed from the same source, because the pattern of gravitational instability in the zone of asthenospheric partial melting tends to persist.

In summary then, a spreading axis is fed by magma produced by partial melting of asthenosphere in an upwelling zone along the axis. Magma supply is discontinuous along the axis, on a scale of a few tens of kilometres. Each magma chamber or crystal mush zone (limited in length by the scale of this magmatic segmentation) may supply lava to several volcanic vents situated every few kilometres along the axis.

An offset of more than about 30 km almost always corresponds to a transform fault, defining the end of a first-order segment. Smaller offsets that define the ends of second- and third-order segments are typically not transform faults.

4.2.4 CHANGES IN SPREADING PATTERN

We have discussed how spreading axes vary along their length, and how overlapping spreading centres may evolve over a time-scale of 100 000 years. What about longer-term changes?

Magnetic anomaly patterns and other data sometimes show that a first-order segment of ridge axis has jumped sideways by several tens or even a few hundreds of kilometres. You should be able to find several places on Figure 3.6 where the amount of displacement between ocean floor age strips on either side of a fracture zone has changed over time, indicating that ridge jumps have occurred. In some instances, a major length of ridge (comprising one or more complete first-order segments) has relocated in what appears to have been a single event. In others, the 'jump' has occurred more gradually by the propagation of the end of one segment at the expense of its less vigorous neighbour.

The anomalously wide age strips to the east of the East Pacific Rise between 10° S and 35° S are probably due to a large ridge jump. In this case, the old ridge is still visible as a bathymetric feature between the active East Pacific Rise and the coast of northern South America (Figure 1.11).

Spreading axes must also adjust whenever the direction of plate motion changes. Sharp changes in spreading direction of up to about 10° at a time occur every few million years and are an inevitable geometric consequence of having several plates jostling against one another on the surface of a (virtually) spherical Earth. This is documented by the changes in direction of fracture zones and sea-floor age strips. There are several examples of this in the Pacific, in particular.

4.2.5 CRUSTAL ABNORMALITIES

Fully developed normal igneous oceanic crust (seismic layers 2 and 3) that formed at a major ridge axis is 7 km thick on average. The relative proportions of different crustal layers generally approximate closely to what you saw in Figure 4.1. However, we have seen that amagmatic extension (Section 4.2.2) can result in the crust being locally thin, or even absent, in some segments, particularly on slow-spreading ridges. Oceanic crust that formed at spreading axes in back-arc basins (cf. Figure 3.8) has the same structure as normal oceanic crust, but is generally 2–3 km thinner.

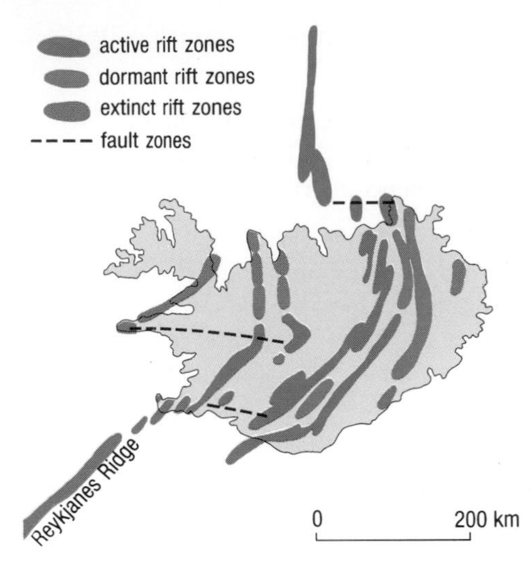

Figure 4.24 Map of Iceland showing its position on the Mid-Atlantic Ridge and the distribution of rift zones. Note how the spreading axis (defined by rift zones) has changed position with time. The Reykjanes Ridge is part of the Mid-Atlantic Ridge system.

By contrast, in some areas the oceanic crust is abnormally thick. The best-known example is the island of Iceland which lies on the Mid-Atlantic Ridge. Here, the processes of crustal generation can be seen on land. A regular spacing of volcanic centres can be seen here, similar to the style of second- and third-order segmentation on the rest of the Mid-Atlantic Ridge, and lateral shifts of the main spreading axis are well documented (Figure 4.24). Iceland, and the shallow part of the Mid-Atlantic Ridge immediately south-west of it (named the Reykjanes Ridge on Figure 4.24), is quite atypical of the Mid-Atlantic Ridge in general. The abnormally thick crust indicates an abundant magma supply consistent with Iceland's location above a mantle plume or 'hot spot'.

Other large (but older and now off-axis) tracts of crust that are oceanic in structure but approaching the thickness of continental crust have been found in locations throughout the oceans (Figure 4.25). The Ontong–Java Plateau (north-east of Papua New Guinea) and the Kerguelen Plateau (which you met in Section 2.6) are among the largest. They seem to be the oceanic equivalent of continental flood basalts, for example the famous Deccan Traps basalts in India. Their upper parts consist of extensive sheets of basaltic lava, and they are understood to be the result of short-lived episodes (less than a few Ma) of greatly enhanced melting at the tops of plumes rising from deep within the mantle. Continental flood basalts and their oceanic equivalents are jointly referred to as **large igneous provinces**. Some, such as the Kerguelen Plateau and the Deccan Traps, are related to episodes of major volcanism during continental break-up and ocean formation. The volcanism heralding the birth of the Red Sea (Section 3.2.1) is a minor example of this. The Ontong–Java Plateau cannot be related closely to continental break-up. It was formed about 122 Ma ago, over a period of between 0.5 and 3 million years, and its enormous volume makes it a feature of global importance.

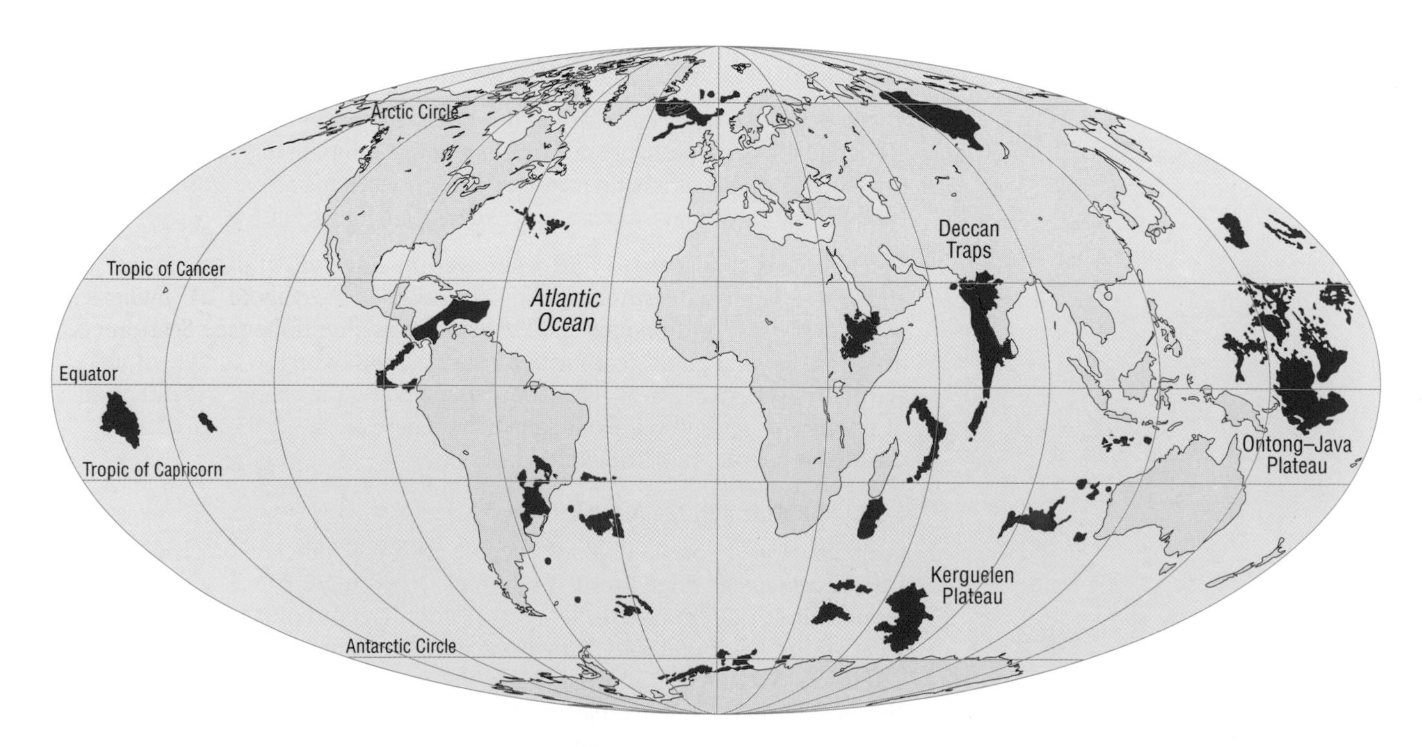

Figure 4.25 The global distribution of large igneous provinces.

QUESTION 4.7 Because its internal structure is poorly known, it is impossible to estimate the crustal volume of the Ontong–Java Plateau precisely. However, it is believed to be between 36 million and 76 million km^3. What are the maximum and minimum rates (in km^3 per year) at which this large igneous province was formed, and how do these compare with the average rate of ocean crust formation for the entire globe, which (over the past 150 Ma) has been 16–26 km^3 per year?

Such spatially and temporally concentrated high rates of volcanic activity would probably have affected the chemistry of seawater (because of reactions between hot rock and seawater, of the sort you will meet in the next Chapter) and could have influenced the climate (through both heating of the oceans and liberation of gases into the atmosphere).

Turning to more normal areas of ocean floor, there is considerable variation in seismic layer 1, which does not everywhere consist exclusively of sediments. Drilling projects such as ODP and DSDP (Section 1.3) have revealed that igneous rocks occur quite commonly within the sedimentary sequence. We know that volcanism forming seamounts and guyots can be much younger than the underlying igneous rocks upon which they are built (Section 2.5.2). It follows that layers of lava or sheet-like basaltic intrusions (sills) may be interlayered with the sediments.

Where there is much faulting and fracturing, as in axial rifts and along transform faults and fracture zones, deeper layers of normal oceanic crust can be exposed at the sea-bed. The greatest vertical displacements are typically at ridge–transform intersections (Figure 2.16).

Indeed, many samples of the full range of rock types representing the four layers shown in Figure 4.1 had been dredged from the ocean floors long before the realities of sea-floor spreading were recognized, and provided key evidence for our understanding of oceanic crustal structure. Nowadays, submersible and deep-tow camera surveys, as well as drilling operations, frequently reveal rocks typical of layer 3 (gabbro) and layer 4 (peridotite) in the walls of median valleys or the scarps and cliffs of transform faults. Peridotite may also be exposed directly as a consequence of amagmatic extension (Section 4.2.2).

The interaction of heated seawater with peridotite of layer 4 alters it into a rock known as **serpentinite**, which is much less dense than peridotite and more plastic, so it can be squeezed up towards the surface along fractures in the crust. This occurs especially at 'leaky' transforms (Figure 2.17) where it is common to find extrusions and intrusions of serpentinite (see Section 5.2.1).

In short, although the structure of oceanic crust is very simple and straightforward in general, it can be complicated in detail locally. These local complications are now easier to detect and investigate than they were formerly, because (as noted in Section 1.3) sophisticated geophysical instruments of the kind widely used in the oil industry can be lowered down boreholes to measure the properties of the rocks. Parameters such as fracture spacing, clay mineral content, seismic velocity and degree of metamorphic alteration can all be measured or deduced, and direct observation is possible with down-hole video cameras. Careful calibration of the instruments, using cores obtained from selected reference boreholes, has made it possible sometimes to dispense with the very expensive business of obtaining rock cores from every borehole drilled into igneous oceanic crust.

So far, we have concentrated mainly on processes that occur at ocean ridges. These are the most active volcanic regions on Earth, but they are not the only regions of volcanism in the ocean basins.

4.3 SEAMOUNTS AND VOLCANIC ISLANDS

A glance at Figure 1.11 shows that seamounts and volcanic islands are a ubiquitous feature of the oceans, though they are more abundant in some parts (e.g. the Pacific) than others (e.g. the Atlantic). Submarine volcanoes have shapes that are generally similar to those of some land-based volcanoes (Figure 2.21), and they grow in much the same way – by successive additions of lava (in this case, mainly pillow lavas) and volcanic ash, which is finely fragmented material produced in explosive eruptions.

Volcanic ash is rare in the oceanic crust proper, because water depths of a few kilometres exert hydrostatic pressures in the order of a few hundred atmospheres, which is enough to prevent dissolved gases escaping from lava. However, lava may break up into rubbly or glassy fragments as noted in Section 4.1.1, and at depths of less than about 500 m gas pressure may overcome the hydrostatic pressure, in which case the rapid escape of dissolved gases can completely fragment the lava into a fine ash composed of glassy shards. The lower parts of oceanic volcanoes are therefore made up largely of pillow lavas, but in volcanoes that have been sufficiently close to sea-level to allow gaseous fragmentation of the lava the upper parts consist of lavas interbedded with ash.

At shallow depths, the chances of explosive eruption and fragmentation of lava are also enhanced when seawater comes into contact with magma or rock still hot enough to boil it, in which case the resulting sudden expansion of large volumes of steam may produce spectacular explosions (Figure 4.26).

Most seamounts are volcanoes that never grew to sea-level; others stopped erupting above sea-level and were subsequently eroded.

What explanation have we offered for the flat tops of guyots that may now be hundreds of metres below sea-level?

Figure 4.26 The explosive eruption of Surtsey (south-west of Iceland), November 1963. When submarine volcanoes build up to near sea-level, and seawater comes into contact with hot magma, the confining pressures are no longer sufficient to prevent explosive conversion to steam. After the volcano has built up far enough above sea-level for the magma column to be isolated from seawater, this type of eruption ceases. Volcanoes commonly explode for other reasons too, such as the build up of gas pressure in the magma.

They were planed off near sea-level by the erosive action of waves (Section 2.5.2) and have since subsided along with the cooling lithospheric plate beneath them (cf. Figure 2.13).

However, the flat tops of some guyots could be congenital rather than developed after formation of the volcano. Figure 4.27(a) shows a flat-topped volcano in the Afar region of Ethiopia, at the southern end of the Red Sea, thought to have been formed below sea-level and subsequently uplifted. Figure 4.27(b) shows a steep-sided and relatively flat-topped seamount in the eastern Pacific. Swath bathymetry and side-scan sonar techniques show that the 'flat' top is due to the presence of a large, shallow crater.

(a)

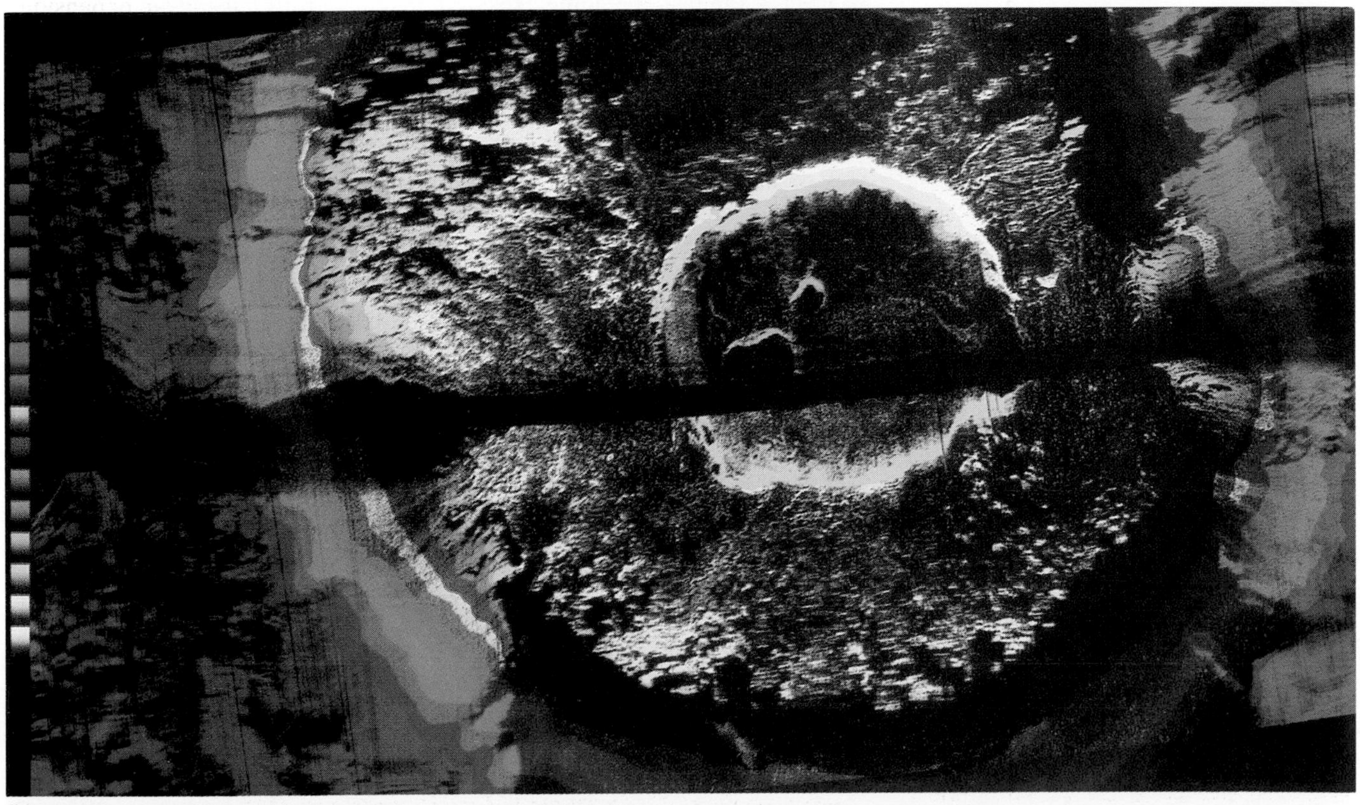

(b)

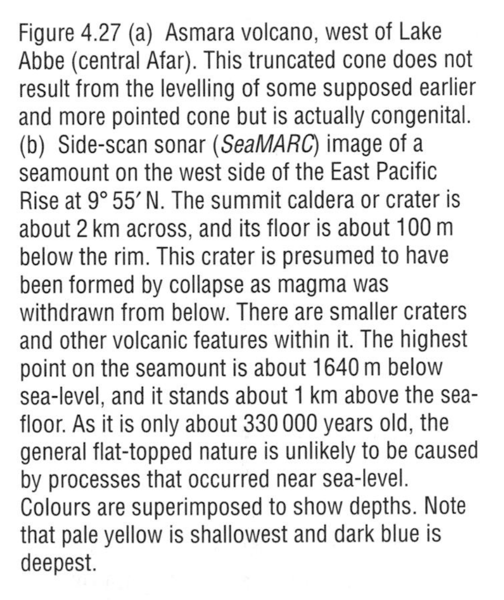

Figure 4.27 (a) Asmara volcano, west of Lake Abbe (central Afar). This truncated cone does not result from the levelling of some supposed earlier and more pointed cone but is actually congenital. (b) Side-scan sonar (*SeaMARC*) image of a seamount on the west side of the East Pacific Rise at 9° 55′ N. The summit caldera or crater is about 2 km across, and its floor is about 100 m below the rim. This crater is presumed to have been formed by collapse as magma was withdrawn from below. There are smaller craters and other volcanic features within it. The highest point on the seamount is about 1640 m below sea-level, and it stands about 1 km above the sea-floor. As it is only about 330 000 years old, the general flat-topped nature is unlikely to be caused by processes that occurred near sea-level. Colours are superimposed to show depths. Note that pale yellow is shallowest and dark blue is deepest.

This has been a very important Chapter, in that it has described in some detail the various processes by which the ocean floor is thought to be created, and has explored some of the evidence upon which the relevant theories are based. It has also provided you with the background you need to study the processes of hydrothermal circulation which occur mainly at ocean-ridge crests – the subject of the next Chapter.

4.4 SUMMARY OF CHAPTER 4

1 In most places, oceanic crust has a seismically well-defined layered structure. Layer 1 consists mostly of sediments overlying the igneous crust of layers 2 and 3. On the simplest interpretation, layer 2 is volcanic, dominated by pillow lavas and other types of basaltic lava (above) and basaltic dykes (below), and layer 3 is gabbro and represents magma that has crystallized at depth. However, the layer 2/layer 3 boundary is sometimes found to represent an increase in the intensity ('grade') of metamorphism with depth. Layer 4 is the uppermost (lithospheric) mantle. The total thickness of normal igneous oceanic crust is about 7 km. In both chemical and mineralogical composition, virtually all the rocks of layers 2 and 3 are basaltic. Layer 4 is peridotite in composition. The principal seismic discontinuities in the oceanic crust are at the base of layer 1 and the top of layer 4 (the Moho). Seismic velocities generally increase with depth, and variations in the gradient of this velocity change enable subdivisions of the crust to be recognized.

2 Seismic tomography and seismic reflection studies have shown that spreading axes are underlain by zones of crystal mush, at a depth equivalent to layer 3. On fast-spreading axes, thin magma lenses have been identified at the top of this zone, but slow-spreading axes generally lack persistent magma bodies.

3 Several sophisticated techniques – including swath bathymetry, side-scan sonar, underwater photography, and submersible operations – have shown that volcanism in the median rift valley of ocean ridges is not continuous but episodic, at intervals of around 10^4–10^5 years. Lavas are erupted from volcanic abyssal hills and hummocky volcanic ridges a few kilometres long, rather than from continuous fissures.

4 Spreading axes are segmented on a variety of scales. Transform faults define the ends of first-order segments, which may be hundreds of kilometres in length. Second-order segment boundaries are not transform faults, but are persistent enough to leave mappable bathymetric traces off-axis; on fast-spreading axes these are called overlapping spreading centres, and on slow-spreading axes they are termed non-transform offsets. Third-order segments are bounded by deviations from axial linearity (devals). Some third-order and all second- and first-order segment boundaries occur where there is a diminution or cessation in magma supply. This occurs about every 10 km or so. Ridges are highest, and the crust thickest, at the midpoints of segments, where the volcanic activity is greatest. This falls off towards the ends of segments, where the ridge is topographically lower, and the crust thinner. The gabbro of layer 3 sometimes thins out to nothing close to fracture zones, and virtually the entire crust may be missing where amagmatic extension occurs.

5 As a model for axial processes, it seems that an elongate zone of less dense asthenosphere rising beneath denser lithosphere will form regularly spaced protrusions into the overlying layer. Each of these protrusions feeds a crystal mush zone or small magma chamber beneath a segment of spreading axis (except for those spreading amagmatically). Each crystal mush zone or magma chamber may feed several volcanic vents in an episodic fashion.

6 Oceanic crust is thinner than normal in back-arc basins and thicker than normal in some other parts of the oceans, notably in large igneous provinces.

7 Fractures and faults can bring rocks of deeper crustal layers to the surface.

8 Seamounts and volcanic islands are formed by isolated submarine volcanoes, building up from the sea-bed. Only a minority reach the surface to form islands. Flat-topped seamounts (guyots) are mostly islands that have been planed off by wave action, but some may have formed with flattish tops due to volcanic processes. Dissolved gases cannot escape from lava at depths of more than about 500 m, so volcanic ash is rare in the deep oceans.

Now try the following questions to consolidate your understanding of this Chapter.

QUESTION 4.8

(a) Overlapping spreading centres have been described in the literature as: 'complex and variable and inherently unstable on length- and time-scales of about 10 km and 10^5–10^6 years respectively, though on longer length- and time-scales they appear to be stable'. Explain what this means.

(b) Look at Figure 4.19(b), which shows both 'limbs' of the overlapping spreading centre at around 9° N. From which magma lens would you expect rock samples dredged from each limb on the profile at 9° 5′ N to have come, and would you expect the samples to be similar or different in terms of their chemical and mineralogical composition?

QUESTION 4.9 Although some dissolved gas generally escapes from any lava, often some gas remains trapped in the rock in small, bubble-like cavities called *vesicles*, which are generally less than 1 cm in diameter. Other things being equal, would you expect vesicles in submarine lavas to be larger or smaller than those in terrestrial lavas? How does your answer help to explain why pillow lavas predominate over volcanic ash in the lower parts of oceanic volcanoes?

CHAPTER 5 HYDROTHERMAL CIRCULATION IN OCEANIC CRUST

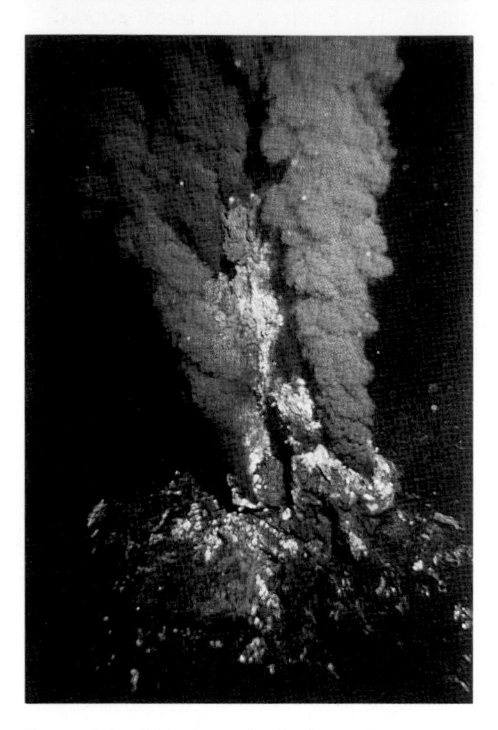

Figure 5.1 A black smoker in the axial zone of the East Pacific Rise. Heated and chemically changed seawater emerges from the sea-bed as clear fluid at 350 °C or more, and immediately precipitates metal sulphide particles on contact with cold bottom water, building the vent chimney and forming the dense plume of black 'smoke'. The vent is about 20 cm across.

The discovery of hot springs on ocean-floor spreading axes during the 1970s was one of the most exciting events in the history of oceanography. The most spectacular are the **black smokers** (Figure 5.1) where hot water gushes out of the vents in the sea-bed at temperatures of about 350–400 °C, forming a dense plume of black 'smoke' made up of minute particles of metal sulphides. At lower vent temperatures (below about 330 °C), any 'smoke' is usually dominated by white particles of barium and calcium sulphate ($BaSO_4$ and $CaSO_4$), giving rise to the term **white smokers**. Less spectacular but just as significant are the lower temperature **warm-water vents** (sometimes known as diffuse vents) where water emerges at about 10–25 °C. The temperature of the surrounding bottom water is about 1–3 °C, and although the fluid escaping from warm-water vents is free of 'smoke' particles, and is therefore clear, it is identifiable visually because of a shimmering effect, like heat-haze in the air. Each of these sources of hot water support unusual ecosystems (Figure 5.2), in which the primary production that underpins the local food web depends not on photosynthesis but on **chemosynthesis** by bacteria (and other micro-organisms), which derive their energy from the oxidation of sulphide from the vents. Some of these primary producers live symbiotically within multicellular host organisms, some form bacterial mats coating the sea-bed (cf. Figure 4.15(a)), and others survive actually within vent chimneys.

As soon as the oceanic crust has been formed by igneous activity, **hydrothermal** processes take over, as seawater (driven by convection) circulates within the newly formed hot igneous rocks. This is by no means a minor phenomenon, and it is likely that about one-third of the entire sea-floor has seawater circulating through it, though high-temperature vents are confined to spreading axes and active off-axis seamounts. The rate of circulation is sufficient for every drop of the 1.4 billion (10^9) km^3 of ocean water to pass through the oceanic crust in about ten million years, leading to major exchanges of chemical elements between seawater and hot basalt. Consequently, the crust buffers the chemical composition of the oceans and for some elements in seawater it is a more important source than rivers. The formation of concentrations of metal sulphide which accompanies the discharge of hydrothermal fluids into the ocean is one of the Earth's principal

Figure 5.2 Tubeworms are a major part of the ecosystems round hydrothermal vents in the Pacific Ocean, as shown in this photograph from the Galápagos spreading axis (which lies between the Cocos and Nazca Plates; see Figure 2.2). The worms are red and can be seen protruding from their chitinous tubes, which may reach several metres in length. They have no stomachs, but absorb nourishment from the water with the aid of symbiotic bacteria that live within their bodies. The small crabs in the photo are blind and have no eyes in their eye sockets, which have become adapted into scrapers for scratching off micro-organisms coating the worm tubes, on which they feed. Tubeworms have not been found on the Mid-Atlantic Ridge; around many vents there, the most prominent fauna are specialized species of shrimp. Note that all these animals live in the *cold* bottom waters surrounding the vents.

Figure 5.3 Geyser activity in Iceland. Heated groundwater and steam rise explosively and intermittently from small vents.

mechanisms of ore generation. Sulphide ore deposits in ophiolites such as those in Cyprus and Newfoundland are well-known examples.

It was predicted that hydrothermal circulation must occur in oceanic crust at ridge crests several years before the first vents were discovered. In the mid-1960s, the occurrence of hydrothermal systems in volcanic areas on land led to the proposition that similar systems should also be found along the ocean-ridge system, which had recently been recognized as a zone of active volcanism. The hot springs and geysers of Iceland (Figure 5.3), which straddles the Mid-Atlantic Ridge (Figure 4.24), provided obvious and highly visible evidence that hydrothermal activity could indeed occur at ridge crests. At about the same time, chemical analyses of samples of the most recently deposited sediments on the sea-floor revealed a systematic increase in the concentrations of compounds of iron, manganese and some other metals (e.g. silver (Ag), chromium (Cr), lead (Pb), zinc (Zn)) towards the ridge crests (Figure 5.4). It was clear that local sources must be responsible for this pattern, and hot spring activity at ridge crests provided the most likely explanation for the observed trends.

Figure 5.4 A map (made in the 1960s) showing the proportions of aluminium (Al), iron (Fe) and manganese (Mn) in the uppermost sediments on the ocean floor (where they occur mostly in the form of oxides). The ratio decreases with distance from ridge crests, because Al is a major constituent of the clay minerals that occur in all deep-sea sediments, and are supplied to the oceans mainly from continental weathering; whereas Fe and Mn are deposited primarily as a result of hydrothermal vent activity near spreading axes.

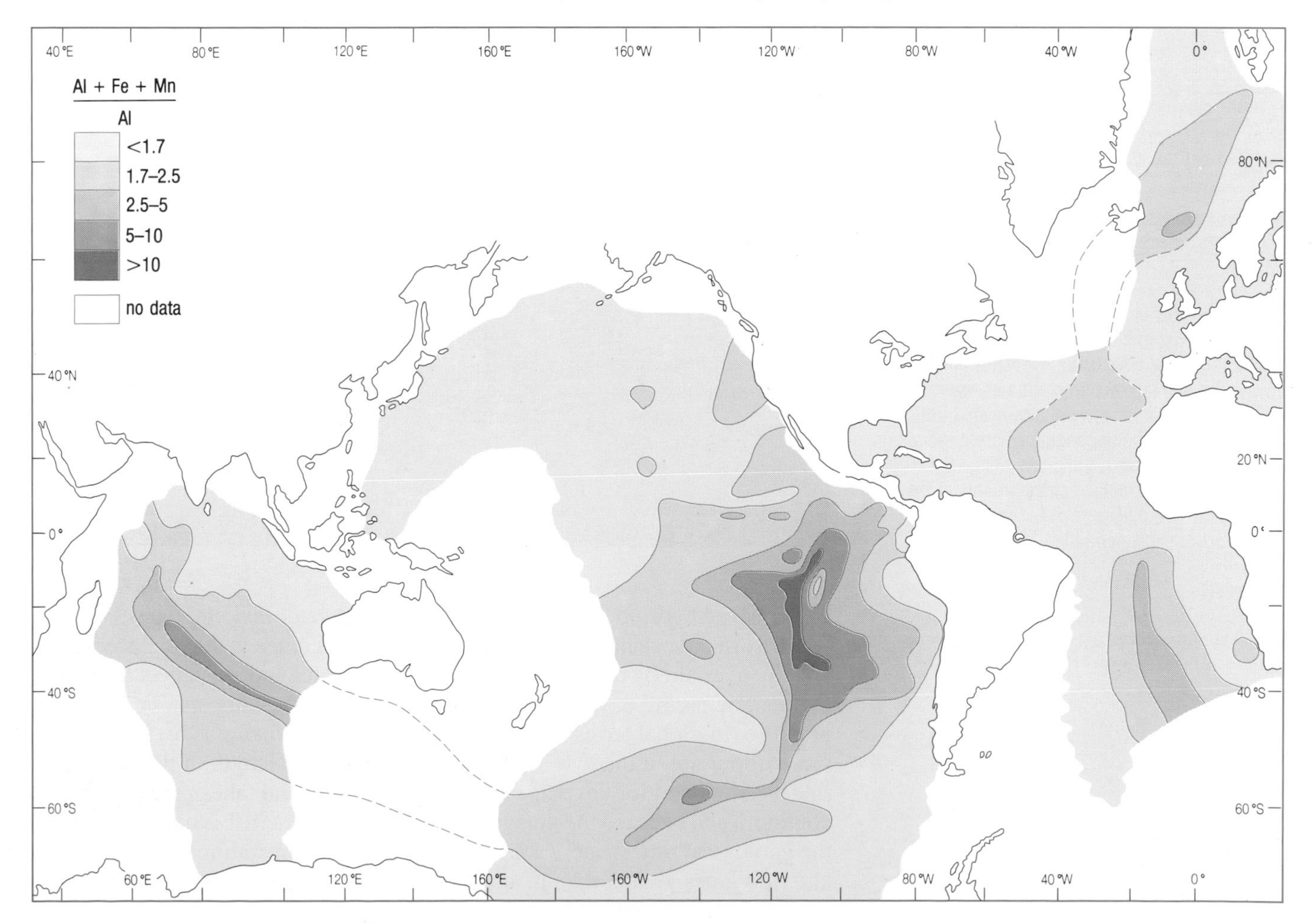

Further support came from samples of basaltic rocks dredged from ridge axes, many of which show clear evidence of having been altered and metamorphosed by reaction with hot seawater. Studies of ophiolites (Sections 3.3.1 and 4.1) and their associated ore deposits confirmed that large volumes of seawater can penetrate more than 5 km into oceanic crust and circulate within it at high temperatures.

5.1 THE NATURE OF HYDROTHERMAL CIRCULATION

Hydrothermal systems in general have two basic characteristics: they occur in regions of high geothermal gradient where hot rocks lie near the surface, and they have a 'plumbing system' of fractures so that cold water can percolate downwards into the crust and then rise convectively to the surface after being heated. The movement of water through these fractures is such that downward percolation occurs slowly over a wide area, through pores, cracks and crevices in the rocks, whereas the upward flow is concentrated through a limited number of channels, which is why it often escapes so vigorously (Figure 5.5).

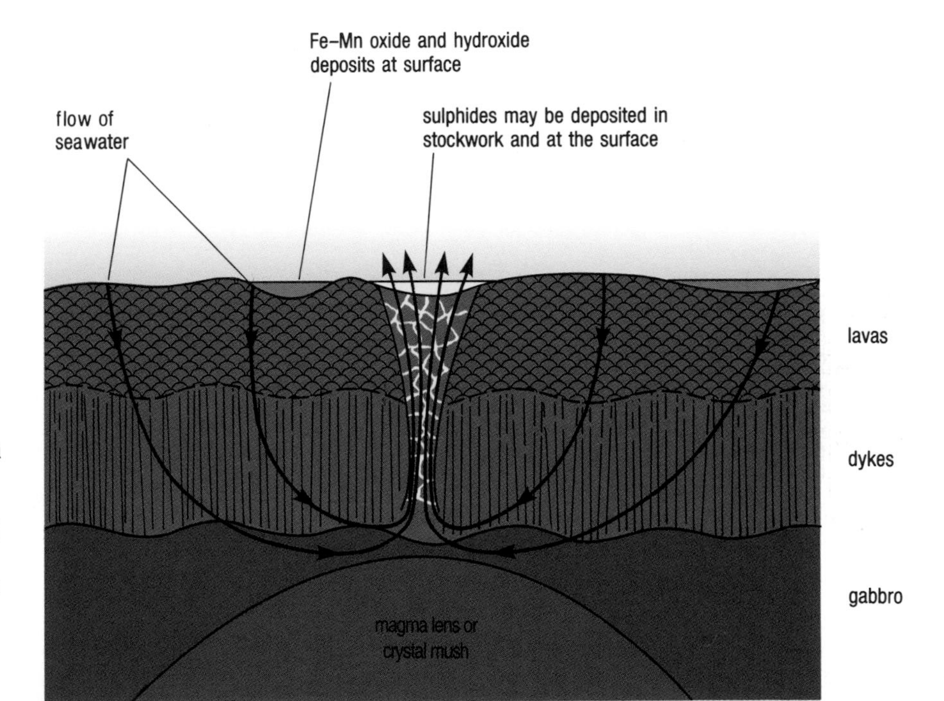

Figure 5.5 Schematic cross-section, normal to a spreading axis, to illustrate hydrothermal convection in oceanic crust. Flow lines indicate downward and lateral percolation of seawater and its expulsion in a narrow zone of heated upflow. This can lead to the deposition of ore minerals on the surface and within the crust; hence the high concentrations of metal-rich sediments at ridge crests indicated by Figure 5.4. The term *stockwork* is explained later – see Section 5.3.1.

In terrestrial hydrothermal systems, it is groundwater that circulates, whereas in the oceans it is seawater. Another difference is that the oceanic crust is overlain by thousands of metres of seawater and so subject to high hydrostatic pressure, whereas on land there is only the atmospheric pressure. In global terms, oceanic hydrothermal circulation is by far the more important of the two. It occurs along the whole length of the ocean-ridge system and for large distances on either side, and takes place continuously, because the generation of new oceanic crust is continuous on the geological time-scale. Along spreading axes, temperatures are higher and flow rates greater than in any terrestrial hydrothermal system.

5.1.1 HEAT FLOW, CONVECTION AND PERMEABILITY

The oceanic crust is hot at its lower boundary whereas the upper boundary is in contact with seawater or wet sediments at 1–3 °C. The temperature difference is most marked near spreading axes, particularly fast ones, where hot (> 1000 °C) magma lenses lie near the top of the gabbro layer.

Heat must rise from the hot to the cold part of the crust. In the solid crust this can happen in two ways: by conduction, in which heat diffuses upwards; and by convection, in which the heat is transferred by the mass movement of water, which is described as **hydrothermal circulation**. The question is, which of the two is more important?

The temperature difference between the top and bottom of the crust becomes progressively less as the lithosphere cools on moving away from the active ridge crest. You have already seen how thermal contraction of the lithosphere leads to an exponential relationship between age and depth to the sea-floor (Figure 2.13). If heat were transferred within the crust by conduction only, a similar exponential relationship with age would hold for the rate of heat loss from the top of the crust.

Measurements of conductive heat flow (the rate of heat loss by conduction per unit area of surface) are made by embedding sensitive heat-measuring instruments into sediments on top of the igneous crust. Figure 5.6 encapsulates some of the results of both theoretical prediction and actual measurement throughout the ocean basins.

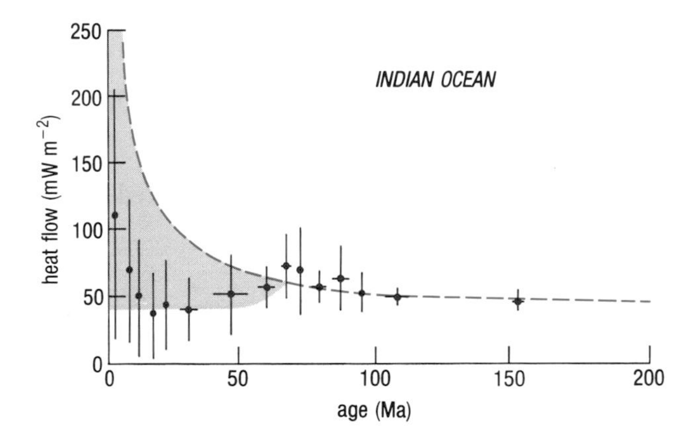

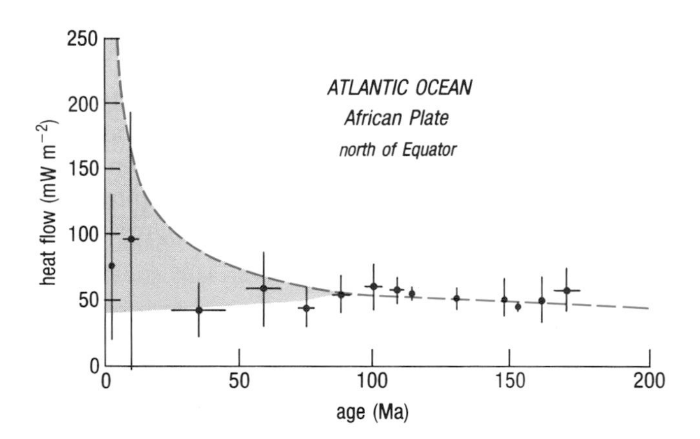

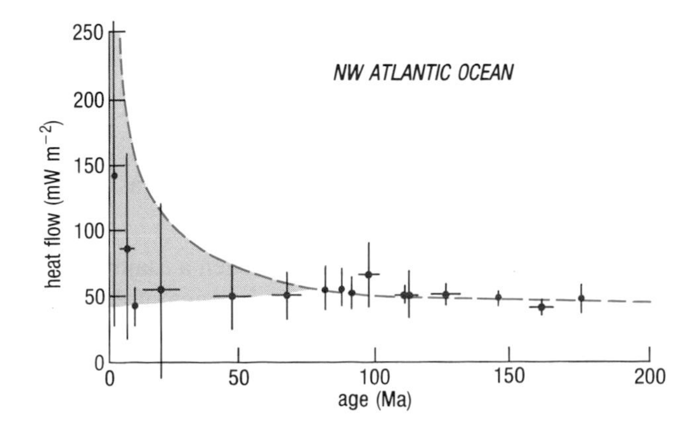

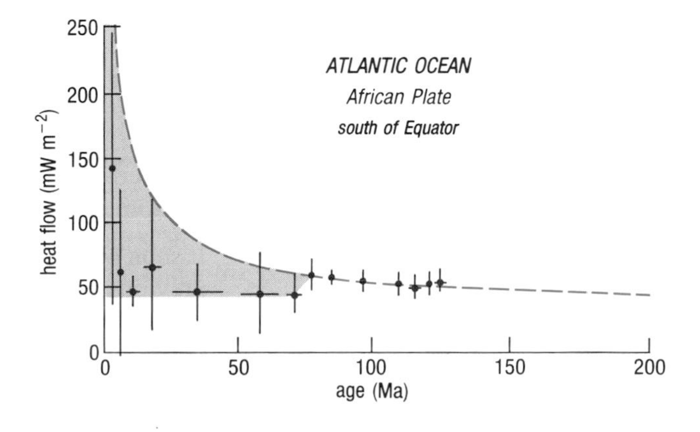

Figure 5.6 Theoretical and observed profiles for heat flow (rate of loss of heat per unit surface area) versus age of oceanic crust for different parts of the world's oceans. Dots at centres of crosses are regional averages of measured heat flow over the age intervals represented by horizontal bars. Vertical bars represent one standard deviation from the mean of the heat flow measurements. Dashed curves represent theoretical exponential decay of heat flow away from mid-ocean ridges, assuming heat loss by conduction only. Shaded areas represent the 'missing' non-conductive heat loss, which cannot be measured by the instruments.

the system, and how fast it moved. Moreover, we can never be sure that vent waters have gone through just a single reaction cycle. For example, greenschist grade rocks consisting almost entirely of quartz and chlorite have been recovered from the sea-bed, with SiO_2 contents of 60–70% or more, and MgO contents of 5–7% or less.

What does this information imply about the exchange of elements between water and rock in this example, compared with the typical exchanges of elements demonstrated by Tables 5.1 and 5.2?

The implication is that silica has been added to the rocks and magnesium has been leached from them, which is the reverse of what seems to happen normally, if Tables 5.1 and 5.2 are representative. Bearing in mind these uncertainties, we should not expect all hot solutions emerging at black smoker vents to have exactly the same composition and temperature as that shown in Table 5.2(b). Some degree of variability is to be expected, though the relative proportions of the principal constituents probably do not change much.

Perhaps the most atypical present-day hydrothermal systems occur in the Red Sea (Figures 3.3 and 3.4), where axial deeps contain thick accumulations of metal-rich muds overlain by concentrated hydrothermal brines at temperatures of 60 °C or more and with salinities of over 300 parts per thousand, an order of magnitude greater than that of normal Red Sea water (itself rather high at nearly 40 parts per thousand). Despite their warmth, the high salinities of these brines makes them denser than normal Red Sea water, so they become ponded within the steep-sided deeps and are unable to escape. (It is possible that their high salinities may in part be due to solution of salts from the nearby evaporites (Section 3.2.1).)

5.3 BLACK SMOKERS – AN EXERCISE IN PREDICTION

One of the more remarkable aspects of the oceanic hydrothermal story is that both the temperature and a close approximation to the composition of high-temperature vent solutions (Table 5.2(b)) were predicted two years before such solutions were sampled.

The first hydrothermal springs to be found in the oceans were located during 1977 in 2500 m of water on the Galápagos spreading axis in the eastern equatorial Pacific at 86° W (between the Cocos and Nazca Plates; Figure 2.2). These are warm-water vents, where water emerges gently at exit temperatures in the range 6–20 °C above the ambient bottom water temperature (2 °C). Analysis of the solutions and comparison with experimental results strongly suggested that these comparatively low-temperature warm springs resulted from simple mixing between a high-temperature hydrothermal solution and the ordinary seawater which saturates the upper parts of seismic layer 2A (Figure 4.1).

Laboratory experiments had demonstrated that magnesium is completely removed from seawater during high-temperature seawater–rock reactions. Therefore, any magnesium in low-temperature springs was likely to originate from mixing of magnesium-free hot hydrothermal water with ordinary seawater. As expected, a negative correlation was found between temperature and concentrations of magnesium Mg^{2+} in samples from low-temperature vents. Extrapolation of this trend to zero Mg^{2+} gave an intercept on the temperature axis at 350 °C, indicating that this was the temperature of the hot hydrothermal component. Similar extrapolations for other constituents allowed predictions to be made about the composition of the high-temperature solutions. The search was then on to find places where this hot component was vented without previous dilution.

Success came in 1979, when black smokers were first seen on the crest of the East Pacific Rise at 21° N, beneath 2500 m of water. Here, solutions at about 350 °C are vented from the sea-floor at speeds of up to several metres per second, through chimneys up to 10 m high, composed mainly of sulphides of iron, copper and zinc. These sulphides are formed by precipitation from solution as the hydrothermal fluid mixes with the surrounding seawater. As mentioned earlier, most of the precipitation occurs as very fine-grained sulphide particles, which form the dense black 'smoke', whereas accumulation of sulphides directly around the vent constructs the chimneys (Figure 5.1). Chemical analyses of the fluid escaping from these vents (Table 5.2(b)) confirmed that the compositional characteristics were as had been predicted from the low-temperature Galápagos vent waters.

5.3.1 BLACK SMOKERS, WHITE SMOKERS AND WARM-WATER VENTS

We have seen that there are two extreme kinds of hydrothermal vent at ridge axes. These are: black smokers, where water emerges at temperatures of about 350–400 °C, and, upon mixing with the bottom seawater, precipitates the mineral particles that build the vent chimney and form black 'smoke' (Figure 5.1); and the more common warm-water vents where the temperature of the clear but shimmering water is rarely more than 35 °C as it emerges, and often less.

Between the two end-members there is a continuum of vent types, represented by 'white smokers', which have exit temperatures in the range approximately 30–330 °C. At these, the vented water may be clear, but it often produces white precipitates, dominated by sulphates of barium and calcium (barite $BaSO_4$ and anhydrite $CaSO_4$), and silica (SiO_2), but with relatively little in the way of the metal sulphides that characterize black smokers (these have presumably already been precipitated *within* the crust, as the rising fluids mix with seawater). Instead, the precipitated particles form white 'smoke', and may build up around the vent to form a pale-coloured chimney.

Table 5.2(b) shows the concentration of barium (Ba^{2+}) in a typical hydrothermal solution to be about two orders of magnitude less than that of calcium (Ca^{2+}), yet you have just read that barium sulphate is abundantly precipitated alongside calcium sulphate at white smokers. Can you suggest why that might be?

The answer is simply that barium sulphate is much more insoluble than calcium sulphate, and so precipitates readily when the vent solutions mix with normal seawater, even at comparatively low concentrations of dissolved barium.

Figure 5.7 illustrates a possible relationship between black smokers, white smokers and warm-water vents, and shows how these hydrothermal vent systems may evolve with time. It suggests that the transition from warm-water vent to white or black smoker can happen at any stage in the overall evolution of a vent field. Such a transition results simply from precipitation of minerals, which reduces the permeability of the surrounding rock and isolates the upflow zone, so that the rising hot water cannot mix with cold water within the rocks near the surface. As you might expect from the previous discussion, the minerals precipitated within the rock include varieties of silica (SiO_2), anhydrite ($CaSO_4$), barite ($BaSO_4$), calcite ($CaCO_3$) and sulphides of iron, copper and zinc. They accumulate to form a sealed lining around a conduit that eventually reaches the sea-bed and then builds the chimney that characterizes black (or white) smokers. Once the upflow is isolated, the vent waters cannot mix, and so emerge at high temperatures and precipitate their dissolved elements on contact with the seawater; hence the black 'smoke' of sulphide particles (or white 'smoke' of sulphate particles).

It follows that beneath hydrothermal vents there are accumulations of sulphide and sulphate minerals in small veins and pockets. These form a *stockwork* which ramifies through the rocks of seismic layer 2, where the hydrothermal fluids have mixed with the cold seawater saturating the upper part of the crust. A warm-water vent or a white smoker that persists without ever evolving into a black smoker must develop a widely dispersed stockwork, which never becomes sufficiently impermeable to isolate the hot upflow completely.

Where black smokers occur, the stockwork has developed in such a way as to prevent mixing within the permeable layer 2. The result is that the solutions reach the surface largely unmixed, so that the sulphides do not precipitate until the vent fluid emerges and mixes with the bottom seawater. Much of the material will settle on the sea-bed surrounding the vent field, but some particles are dispersed by bottom currents over much wider areas (see Section 5.4.2). As indicated schematically in Figure 5.5, the more widely dispersed particles are commonly not sulphides, but oxides and hydroxides of iron and manganese. These are precipitated when dissolved Fe^{2+} and Mn^{2+} from the hydrothermal solutions are oxidized upon mixing with seawater. Nearer to the vents, however, sulphides predominate, both at the surface (e.g. forming chimneys) and in the stockwork beneath.

The oceanic crust is thus likely to be riddled with pockets of metal sulphide stockworks at extinct vent fields. It is likely that the sulphide deposits in ophiolites were deposited from hydrothermal solutions at spreading axes in ancient oceans.

The relationship between vent types shown in Figure 5.7 is generalized and not always applicable. For example, black smokers and white smokers can coexist within a hundred metres of each other (see Section 5.3.3), in which case it seems that the stockwork is locally patchy enough to allow ingress of seawater near the white smokers. Similarly, some warm-water vents occur in close proximity to 'smoking' vents, and others may represent the waning phase of hydrothermal activity when the heat supply that formerly powered a 'smoking' vent has run down.

A particularly striking feature of Figure 5.7 is the very sharp temperature change below the top half-kilometre or so. This is the uppermost part of the igneous oceanic crust, where the proportion of faulted rocks, broken pillows, and debris is greatest. It corresponds to seismic layer 2A, which is believed to have very high permeability, and thus to be in good hydraulic continuity with the overlying seawater, which effectively saturates this layer. Within this layer (where temperatures are less than about 20 °C), only sea-floor weathering reactions are widespread. Significant metamorphism does not go on at depths of less than about a kilometre below the sea-bed except in the immediate vicinity of hydrothermal conduits.

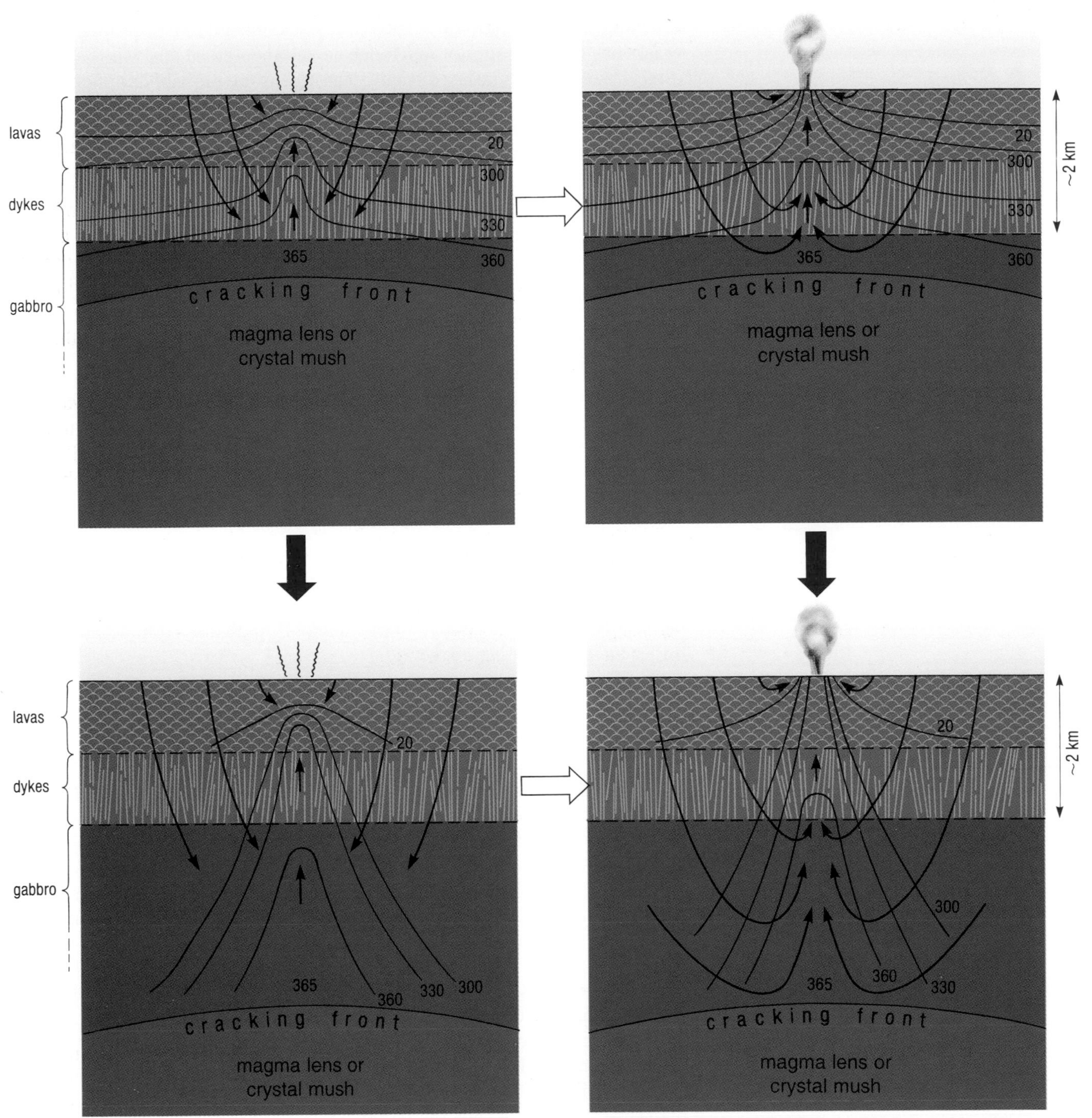

Figure 5.7 The hypothetical evolution of a hydrothermal vent from warm-water vent to black smoker. The depth of penetration of seawater increases with time (solid arrows), and the transition to black smoker (open arrows) can occur at any time. During this transition, the vent is a white smoker. A vent may stop evolving at any stage, and never develop into a black smoker. Numbered lines are isotherms (in °C) for solutions circulating through the rock. The cracking front marks the maximum depth at which cracks occur and below which solutions cannot penetrate. It is approximately coincident with the roof of the magma lens or crystal mush zone, and moves downwards as the magma or crystal mush cools and the hydrothermal circulation is able to penetrate deeper.

5.3.2 THE LIFETIMES OF HYDROTHERMAL SYSTEMS

The depth of circulation of hydrothermal systems increases with time, as the rocks move away from the spreading axis and continue to cool so that cracks are able to penetrate deeper into the crust, as shown by the advance of the **cracking front** in Figure 5.7. The width, spacing and rate of propagation of cracks has been a matter of much debate and theoretical analysis, because direct observation is rather difficult. The cracks are probably no more than 1–3 mm wide, with an average spacing of a few tens of cm up to a metre or two, and the rate of penetration of the cracking front has been estimated to be a few metres per year. Naturally, cracks cannot propagate into a liquid magma body or a crystal mush, and can extend into the gabbro of seismic layer 3 only after it has fully solidified.

Calculations show that the cooling effect of hydrothermal systems (deduced from their rate of heat transport) is probably sufficient to preclude long-term persistence of liquid magma or crystal mush, especially at slow-spreading axes. Thus, in addition to being discontinuous along the axis (Section 4.2.3, Figure 4.23), magma bodies will be episodic wherever hydrothermal cooling is sufficient to completely crystallize the whole gabbro layer at the spreading axis during intervals between replenishment from below.

QUESTION 5.4 So what is the lifespan of axial hydrothermal vents? Estimates have ranged from decades to tens of millennia, but you can make your own assessment. One way is to consider the rate of downward penetration of the cracking front. Another is to consider the hydrothermal heat flux. Try both of these now:

(a) Hydrothermal solutions may penetrate to a depth of at least 5 km, and the rate of advance of the cracking front has been estimated as a few metres per year (say $3\,m\,yr^{-1}$). Where in the range of timescales from decades to millennia does your estimate fall?

(b) A typical rate of heat loss (i.e. *heat flux*) from a single hydrothermal system is about 200 MW (2×10^8 W, where W = watt = 1 joule per second, $J\,s^{-1}$). The volume of magma available for a hydrothermal vent system to extract heat from is approximately 10 km^3 ($10^{10}\,m^3$). The density of gabbro is $2500\,kg\,m^{-3}$, and its latent heat of solidification (the heat given out by gabbro when it solidifies) is $4.5 \times 10^5\,J\,kg^{-1}$. Assuming, for simplicity, that a 200 MW hydrothermal system is powered only by latent heat (thus ignoring heat extracted by *cooling* the solidified gabbro, and any heat input by replenishment of magma), what is the lifetime of the system? (By ignoring these two factors, you will, in effect, be calculating a likely *minimum* lifetime. *NB:* the number of seconds in a year is $\sim 32 \times 10^6$.)

Hydrothermal vent systems have not been studied for long enough to confirm estimates of the type made in answering Question 5.4. The temperature/composition of the fluid emerging at any particular vent have usually been found to be remarkably stable over more than ten years, and small chemical differences between fluids at neighbouring vents are maintained. This is despite the fact that vent chimneys have been seen to grow and fall over, and indicates stability of the circulation system on at least a decadal timescale. However, the exit temperature at what was probably a new hydrothermal vent at 9° 46.5′ N on the East Pacific Rise (created by the same events that produced the 1991 'Tubeworm Barbecue' lava flow (Figure 4.15)) was seen to change significantly over three years (Figure 5.8); changes in H_2S concentration show that the composition of the vent fluid was evolving at the same time.

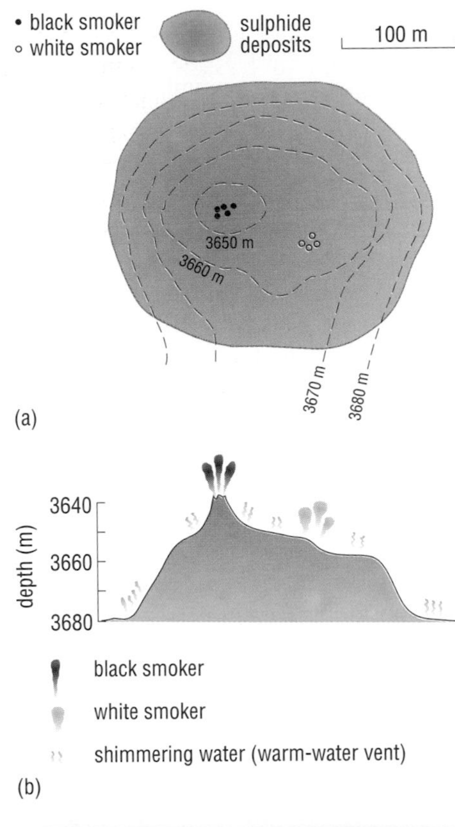

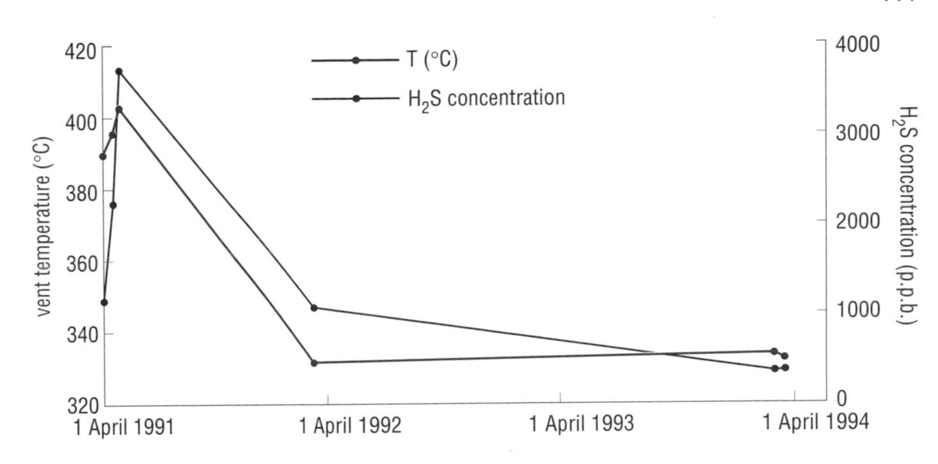

Figure 5.8 Temperature and H_2S concentration of vent fluid at a young high-temperature hydrothermal vent, at 9° 46.5′ N on the East Pacific Rise. (From data in K.L. Von Dam *et al.* (1995) in *Nature*, **375**.)

depth (m)

3640

3660

3680

black smoker

white smoker

shimmering water (warm-water vent)

(b)

(c)

Figure 5.9 The TAG hydrothermal mound, seen in (a) plan view and (b) NW–SE elevation, vertical scale exaggerated. Contours are depth below sea-level. (c) A group of inactive chimneys on the TAG mound, *c.* 1–2 m high. Yellow–brown surfaces are coated with sulphides; bluish-white areas are mostly anhydrite.

5.3.3 ANATOMY OF A VENT FIELD

As more and more hydrothermal vents are discovered and studied, it has become clear that each has its own characteristics. It is rare for a vent to occur in isolation, and usually there are several grouped together in a region a few kilometres across or less (described as a *vent field*). We will look here at one of the larger examples, known as the TAG hydrothermal field (TAG is an acronym for Trans-Atlantic Geotraverse, the name of the survey project that first identified it). The TAG field is in the median valley of the Mid-Atlantic Ridge at 26° 08′ N, within the first-order segment defined by the Kane and Atlantis transform faults examined in Section 4.1.3.

The most prominent feature of the TAG field is a 200 m diameter, 50 m high mound coated by iron and copper sulphides and with an interior that is mostly anhydrite ($CaSO_4$) mixed with sulphides. The mound is topped by a cluster of chimneys discharging fluids at up to 363 °C (Figure 5.9). The mound has a cluster of white smokers on one side, and its steep sides are composed of sulphide debris (talus) that has tumbled from the chimney areas and from which a diffuse flow of shimmering water can be seen.

The TAG mound lies about 2.4 km east of the axis of the median valley, apparently above a fault that offsets its east wall (though not the axis itself). It is estimated that this mound is about 18 000 years old. There are several similar, but extinct, mounds up to 140 000 years old to the east and north, and the east wall of the median valley is a zone of diffuse warm-water venting.

QUESTION 5.5 If the TAG mound had been continuously active since its inception, how would its age compare with the lifetime for a hydrothermal system that you estimated (in two ways) in answering Question 5.4? Can you account for the difference?

In fact, drill cores extracted from the TAG mound provide convincing evidence that its hydrothermal activity has been episodic. During active periods, anhydrite is precipitated within the mound, and sulphides precipitate mainly across the surface. During lulls in the activity, the dome tends to collapse, disrupting its interior into a rubbly mixture of anhydrite and sulphides. The present active episode is responsible for *unbroken* veins of anhydrite that cut through the rubbly interior. Precisely *why* the activity is episodic is not clear (perhaps the deep plumbing keeps becoming blocked), but episodic rather than continuous activity may be the main factor allowing the dome to have had such a long lifetime even if the total heat that it can tap is limited.

Presumably, the TAG mound will become extinct when sea-floor spreading has carried it sufficiently far from the axial zone. This has evidently been the fate of the extinct mounds nearby. The amount of metal sulphides within and below the TAG mound is estimated to be about 4 million tonnes, which is comparable to the size of the ore bodies in the Troodos ophiolite of Cyprus. The TAG deposit is the biggest metal sulphide body yet identified on a spreading axis, with the exception of the metalliferous sediments in the axial deeps of the Red Sea; for example, the Atlantis II Deep (Figure 3.4) contains about 90 million tonnes of sulphides, estimated to contain about 0.5% copper and 2% zinc. In contrast, sulphide deposits associated with vent fields on fast-spreading axes are typically about two orders of magnitude smaller than the TAG deposit. Presumably, faster spreading and a greater rate of magma supply mean that vent fields cannot persist for so long before they are either carried away from the zone of hydrothermal upflow or buried by lava flows.

5.4 THE EXTENT OF HYDROTHERMAL ACTIVITY

The heat-flow versus age relationship in Figure 5.6 and the evidence of metamorphic rocks drilled or dredged from the ocean ridges show that hydrothermal activity must occur (if only intermittently) throughout the entire 50 000 km length of spreading axes in the oceans. Over geological time, the axial zone can be regarded as a permanent linear heat source. Hydrothermal vent fields of the kind we have been describing usually occur close to the ridge axis, usually within the median valley (if present), and they represent the upflow zones of Figures 5.5 and 5.7, above the magma lens or the hottest part of the crystal mush zone. Although the hydrothermal upflow is tightly focused, downflow zones draw in seawater over much wider areas, encompassing the ridge flanks and also significant stretches of the axial zone.

Various off-axis hydrothermal vents have been identified too, in association with off-axis magmatic activity at seamounts. Several hydrothermal ore deposits in ophiolites appear to have been formed in such locations.

Although the mid-ocean ridge system is incompletely explored, it is possible to document the distribution of hydrothermal vent fields fairly comprehensively on some stretches. Through a combination of direct observation of vents and detection of the plumes of effluent emitted by vents (see Section 5.4.2), the Mid-Atlantic Ridge between 20° N and 40° N that we considered in Section 4.1.3 is relatively well known. Here, six distinct vent fields had been identified by the mid-1990s, namely:

Snake Pit (23° 22´ N, in the second-order segment immediately south of the Kane transform fault) with several 335–350 °C black smokers on the crest of a volcanic ridge directly over the axis.

TAG (26° 08´ N, in the centre of the fifth second-order segment north of the Kane transform fault), described above.

Broken Spur (29° 10´ N, in the centre of the second-order segment immediately south of the Atlantis transform fault) with several black smokers at more than 350 °C, situated on axis-parallel faults through the axial volcanic ridge.

Rainbow (36° 15´ N), a large vent field on the non-transform offset between the first and second second-order segments south of the FAMOUS area. It has the greatest rate of discharge of vent fluids yet discovered in the Atlantic.

Lucky Strike (37° 17.5´ N, in the second second-order segment north of the FAMOUS area) where there are five or more black smokers venting fluids at 303–333 °C, and several cooler vents, located between three axial volcanoes.

Menez Gwen (38° 10´ N, in the next second-order segment north from Lucky Strike) with vents distributed over a 200 m^2 area on a large (700 m high, 17 km diameter) abyssal hill.

Other vents are known on the Mid-Atlantic Ridge at 14° 45´ N and just south-west of Iceland. There are also indications of plumes emanating from vents in each second-order segment between Rainbow and Lucky Strike, and further vent-site discoveries are to be expected wherever the Mid-Atlantic Ridge is explored.

Our other case study area from Section 4.1.3 (the East Pacific Rise between 9° and 10° N) has over 40 active high-temperature hydrothermal vents, all within the narrow median valley, and over 100 inactive vents marked by sulphide deposits. Comparable vent fields occur elsewhere on the East Pacific Rise at 11°, 13° and 21° N, and 17°, 18°, 20° and 21° S, and also on the Galápagos spreading axis at 86° W. There is also a field of vents at temperatures ranging from 100 to 315 °C in the Guaymas Basin, a 75-km-long spreading segment half-way up the Gulf of California (near 25° N), which is effectively a northward continuation of the East Pacific Rise.

The only other well-explored spreading axis lies in the north-east of the Pacific (see Figure 2.2). This is most widely known as the Juan de Fuca Ridge (although, confusingly, its northernmost first-order segment is sometimes named the Explorer Ridge and its southernmost first-order segment the Gorda Ridge), and it has ten or more distinct vent fields.

Hydrothermal vents have also been located in several marginal basins in the western Pacific. Among these are the Lau Basin, the North Fiji Basin, the Woodlark Basin, the Manus Basin, the Mariana Trough (the back-arc basin behind the Marianas island arc) and the Okinawa Trough.

Figure 5.10 shows the global distribution of vent fields as known in the mid-1990s. Any such map becomes out of date very quickly because new vent fields are discovered every year. The majority of known vent fields are on relatively shallow parts of ridges (which are easier to explore by submersible), and in areas that are conveniently close to research bases and where good weather prevails. The lack of hydrothermal vents in other areas is probably more apparent than real. We will look at the most effective way of discovering new vent fields (by detecting plumes) in Section 5.4.2, but first we will say a little about vent biology.

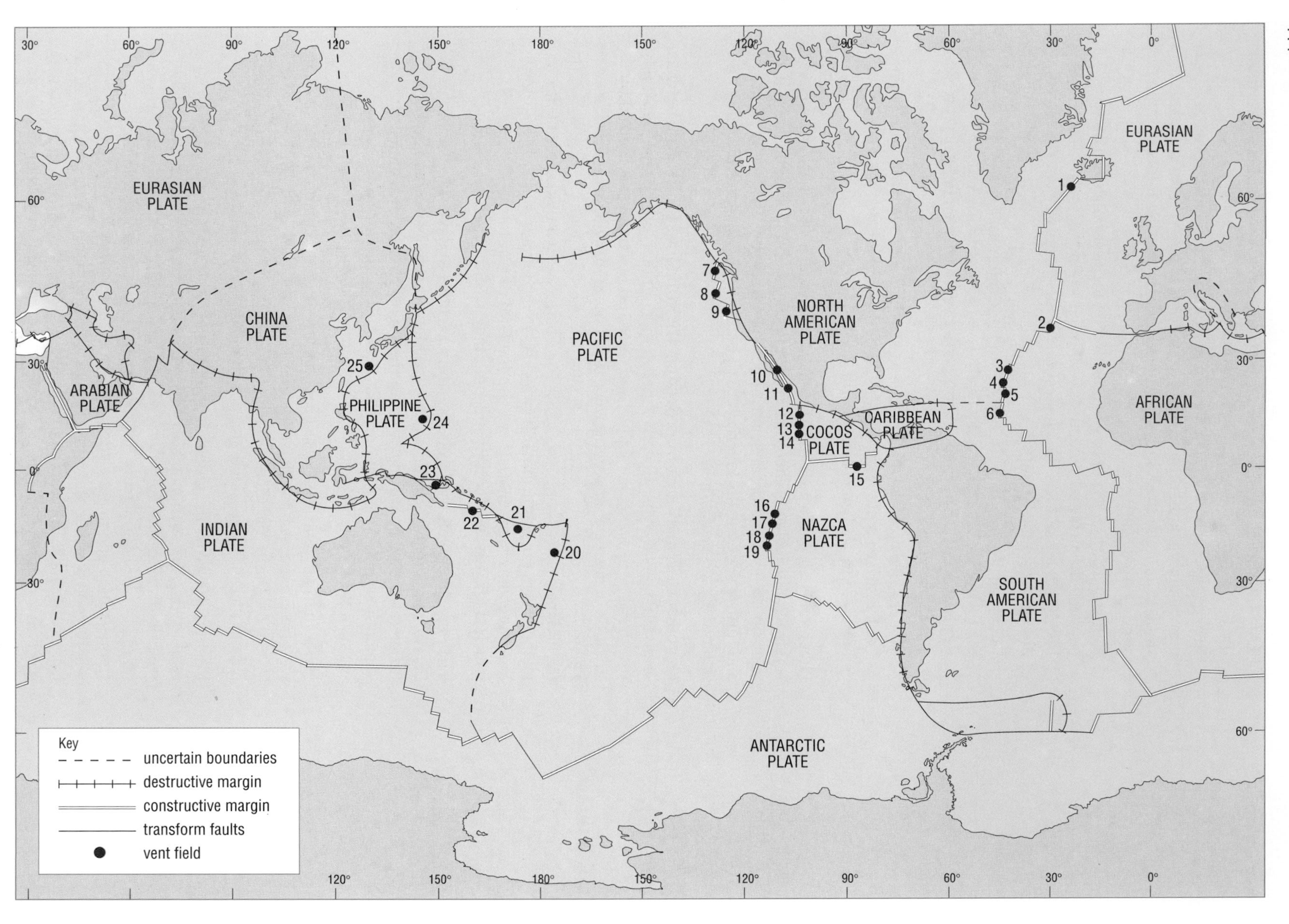

Figure 5.10 The locations of hydrothermal vent fields known in the mid-1990s. Numbers refer to vent fields as follows: 1 Reykjanes Ridge; 2 Menez Gwen, Lucky Strike, and Rainbow; 3 Broken Spur; 4 TAG; 5 Snake Pit; 6 MAR 14° 45´ N; 7 Explorer Ridge vents; 8 Juan de Fuca Ridge vents; 9 Gorda Ridge vents; 10 Guaymas Basin; 11 EPR 21° N; 12 EPR 13° N; 13 EPR 11° N; 14 EPR 10° N; 15 Galápagos Ridge 86° W; 16 EPR 17° S; 17 EPR 18 °S; 18 EPR 20° S; 19 EPR 21° S; 20 Lau Basin; 21 North Fiji Basin; 22 Woodlark Basin; 23 Manus Basin; 24 Mariana Trough; 25 Okinawa Trough.

5.4.1 THE BIOLOGICAL SIGNIFICANCE OF HYDROTHERMAL VENT SYSTEMS

Sections 5.3.2 and 5.3.3 showed that the lifetimes of hydrothermal vent systems are unlikely to exceed tens of thousands of years, and may be less. When an individual vent field finally peters out, the chemical energy source for the vent community ceases, leaving the vent organisms without means of support, so they must die. As the great majority of the adult forms of vent organisms are either sessile (living fixed to a substrate) or slow moving, how can new vents be colonized? The answer appears to lie in the fact that they all produce freely drifting larvae. Only a tiny fraction of these larvae survive, but enough are carried in the right direction by currents to new vents to ensure that the species is established elsewhere by the time an old vent field becomes inactive and its community has died out.

The reliance on larval dispersal to propagate each species is reflected in the fact that, in general, the closer two sites are along a spreading axis, the greater the similarity between their faunas. For example, two vent fields about 800 km apart on the East Pacific Rise have 54 species in common, but the number falls to 11 for sites about 2000 km apart. Just five species are shared between vent fields on the East Pacific Rise and those in marginal basins in the western Pacific, and only one between the Atlantic and the Pacific. It is possible that detailed, systematic analysis of species distributions of vent faunas, and genetic fingerprinting within species, may provide additional evidence to document past configurations of spreading axes and/or deep-sea currents.

Study of the bacteria, and bacteria-like, single-celled organisms that live in the extreme chemical and thermal conditions at vents may prove even more rewarding to science. Although their dispersal mechanisms are uncertain (perhaps individuals can remain dormant for protracted periods), it is thought that similar organisms have lived near hot sea-floor vents throughout the past 3 to 4 billion years and it has been proposed that life may even have originated in such a setting.

5.4.2 HYDROTHERMAL PLUMES

Although a few vent fields have been discovered accidentally during chance dives by submersibles or by cameras carried by devices towed just above the sea-floor, the first indications of most of them have come from detection of the plume of effluent emitted from vents into the water column. Most plumes are detectable over large areas, and only if a plume has been located is it worth searching for the vents themselves. When fluid emerges from a vent it rises for several hundred metres, because of its buoyancy. As it rises, the fluid becomes diluted by mixing with seawater until its density matches that of the surrounding water, so that the plume can no longer rise. It then spreads laterally, becoming diluted still further and dispersed by currents.

Typically, a plume is diluted by a factor of about 10^4 by the time it has risen a hundred metres or so. Despite this high dilution factor, there are several clues that enable it to be distinguished from ordinary seawater at the same depth. Its heat content is one clue; although after dilution the temperature anomaly is only a fraction of a degree, it can be detected using modern instruments. Another good tracer for a plume is the suspended 'smoke' particles that have not yet settled out. The concentration of such particles in the water column can be inferred from the decrease in clarity of the water. This can be measured using either a *transmissometer* (which

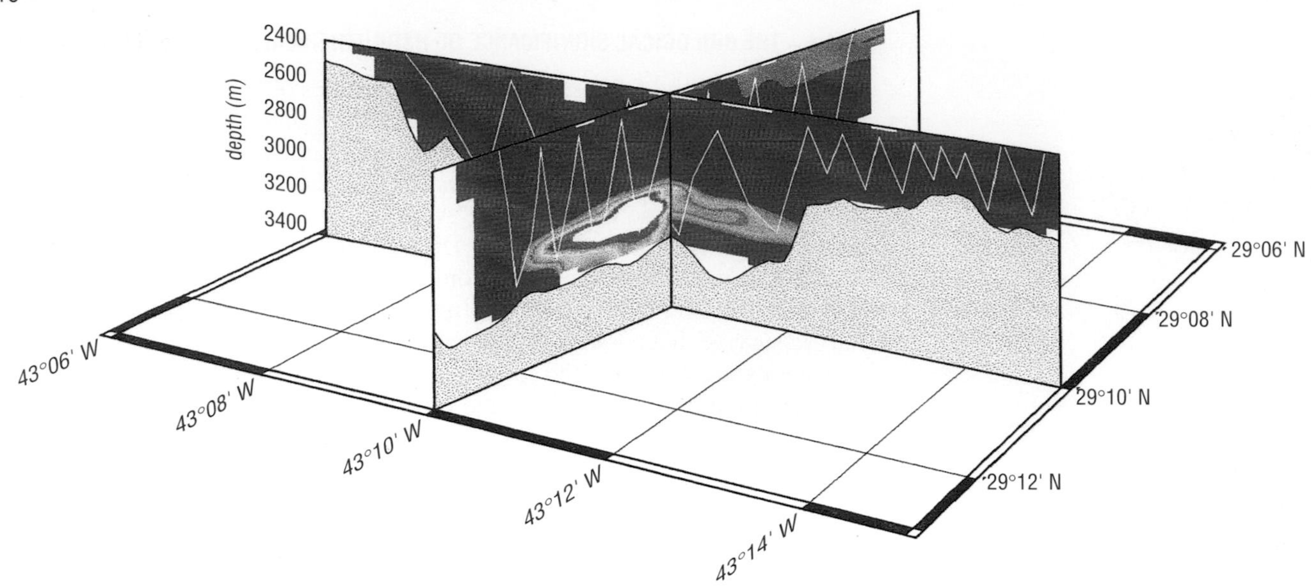

Figure 5.11 Suspended particle concentration used to delineate a hydrothermal plume above the Broken Spur vent field on the Mid-Atlantic Ridge (29° 10′ N). The view is towards the SSE, and shows E–W and N–S cross-sections. The sea-bed (and everything below it) is shaded brown, revealing the topography of the axial valley in the E–W cross-section. Increasing particle concentration is shown by dark blue (clear water) through green, yellow and red to white (highest particle concentration). The particle concentration was recorded by a nephelometer on a deep-towed instrument platform called BRIDGET, operated in so-called 'tow-yo' mode, following an up-and-down zig-zag path. N–S and E–W tow-yo paths are indicated by fine white lines. (Data gaps in the water column are white, and should *not* be confused with high particle concentration.)

measures the proportion of light reaching a detector aimed at a nearby light source) or a *nephelometer* (which measures the amount of light scattered in various directions by the suspended particles). Figure 5.11 shows a plume mapped by a deep-towed nephelometer.

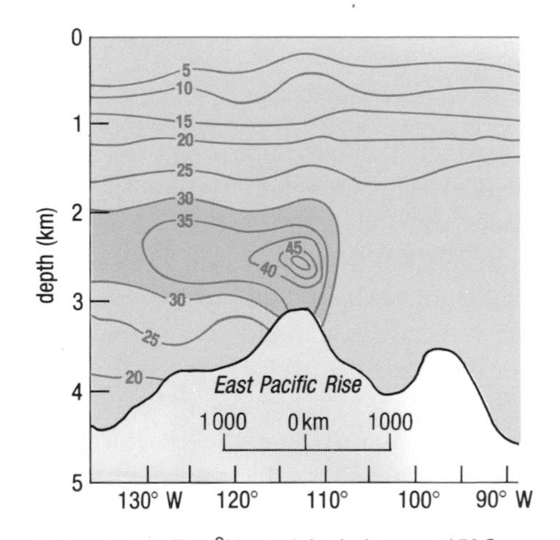

Figure 5.12 The ^{3}He-enriched plume at 15° S emanating from the fastest-spreading part of the East Pacific Rise, showing the marked westward transport by currents at mid-depths. The contours represent ^{3}He : ^{4}He ratios expressed as delta-^{3}He, or δ^3He, which is a measure of the enrichment of ^{3}He in the water relative to the atmosphere. The data plotted here were collected in 1972, but the persistence of the plume was confirmed by a further survey in 1987. The source of the plume is the vent fields near 17° and 18° S, probably supplemented by diffuse venting from unidentified sites.

The most persistent tracers for hydrothermal plumes are dissolved gases, because unlike suspended particles these will not settle out. Dissolved gases emitted at hydrothermal vents include methane (CH_4), hydrogen (H_2), hydrogen sulphide (H_2S), carbon dioxide (CO_2), carbon monoxide (CO), nitrous oxide (N_2O), and helium (He). Although its concentration is several orders of magnitude less than those of methane and hydrogen, helium has proved to be a more potent tracer because of its distinctive isotopic signature.

Helium is a light gas that is rare in the Earth's atmosphere, because it escapes very easily to space. It has two stable isotopes. The commonest is helium-4 (^{4}He), which is produced by radioactive decay of uranium and thorium, both of which are widespread (albeit in trace amounts) in the sediments and igneous rocks of the Earth's crust. The much rarer isotope, helium-3 (^{3}He), has two sources: one is formation by cosmic ray bombardment in the atmosphere, from which it escapes rapidly to space; the other is escape of primordial helium (rich in ^{3}He) that was trapped within the Earth's mantle at the time of its formation. This primordial ^{3}He escapes by outgassing in volcanic activity, including that at oceanic spreading axes.

The ratio of ^{3}He to ^{4}He is far higher in hydrothermal vent waters than anywhere else, and ^{3}He has the additional advantage of being the only constituent of vent solutions that is unequivocally known to have originated in the mantle and not through water–rock reactions or microbial (bacterial) activity. Figure 5.12 illustrates how the dispersal of the hydrothermal plume

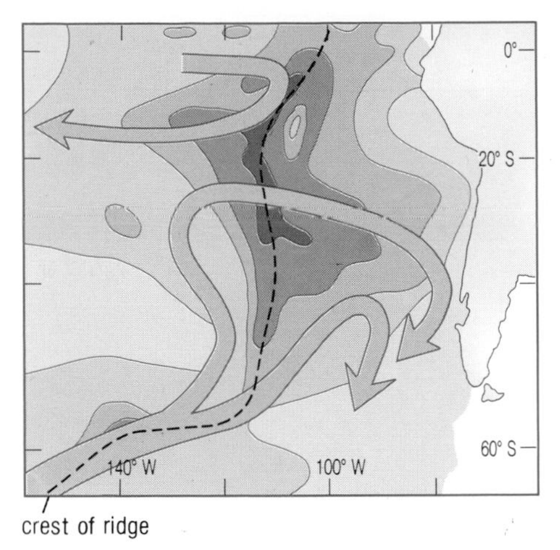

Figure 5.13 A map of the (Al+Fe+Mn)/Al ratio extracted from Figure 5.4 with independently determined current flow (arrows) at ridge-crest depths near the East Pacific Rise south of the Equator. The dashed line is the ridge crest.

above the East Pacific Rise at 15° S can be mapped using the $^3He : ^4He$ ratio. There is a pronounced westward flow at 2–3 km depth and the plume extends thousands of kilometres to the west. Spreading rates are greater along this stretch of the East Pacific Rise than anywhere else in the oceans (cf. Chapter 3) and the ratio of ^3He to ^4He is about ten times higher than above the 21° N vent field. Only in the Red Sea and the Guaymas Basin, which are young deep linear basins with restricted circulation, have higher concentrations of ^3He been found.

If you look back to Figure 5.4, you can see an obvious asymmetry in the distribution of metal-rich sediments about this part of the East Pacific Rise (between 5° S and 40° S). Figure 5.13 shows the independently inferred current regime *at ridge-crest depths* superimposed on this sediment pattern. The correlation seems obvious enough, but before the ^3He-rich plume in Figure 5.12 was observed, circulation models for the Pacific did not include this flow regime. However, independent corroboration of this circulation pattern has now been obtained, using standard oceanic variables (temperature, salinity, density, oxygen, nutrients) which you can read about elsewhere in this Series.

Methane, hydrogen and carbon dioxide can be liberated by partial melting in the mantle (like ^3He) and they are all well-documented constituents of volcanic gases elsewhere. However, unlike ^3He, they can also be released into solution by oxidation–reduction reactions within the basalt–seawater system during hydrothermal circulation, in some cases (notably that of methane) mediated by microbial (bacterial) activity. Such reactions are the dominant sources of other vent gases. Methane is the most abundant gas in vent waters, where $CH_4 : ^3He$ ratios are in the order of $10^7 : 1$. Variations in this ratio may result from different proportions of abiotic (meaning inorganic in origin) and microbial methane, but there may be other causes. For example, methane-enriched plumes of water found in the Mariana Trough marginal basin are not accompanied by the marked ^3He enrichment that characterizes hydrothermal vent fields in the major oceans. This may mean that the upper mantle there has already been depleted in ^3He.

As plumes are easier to locate than the vents themselves, the abundance of plumes above a spreading axis is the best indicator we have of the extent of active hydrothermal venting. Figure 5.14 is a diagram showing plume incidence (defined as the percentage length of axis overlain by a plume) plotted against spreading rate.

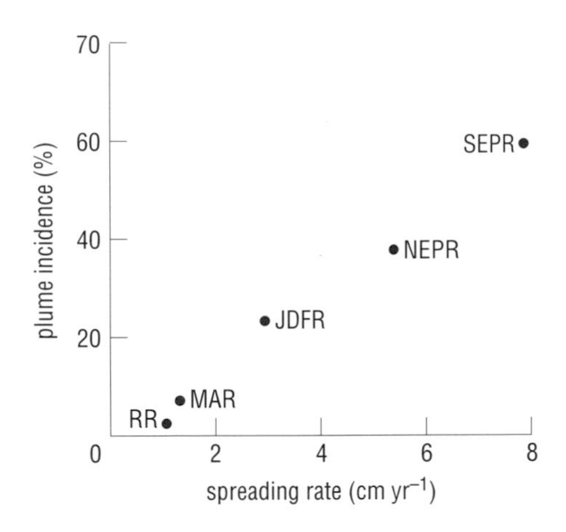

Figure 5.14 Incidence of hydrothermal plumes, plotted against spreading rate. RR = Reykjanes Ridge (the Mid-Atlantic Ridge SW of Iceland); MAR = Mid-Atlantic Ridge (11°–40° N); JDFR = Juan de Fuca Ridge; NEPR = North East Pacific Rise (8° 40′–13° 10′ N); SEPR = South East Pacific Rise (13° 50′–18° 40′ S).

What can you infer about the relationship between spreading rate and plume incidence on Figure 5.14, and is it what you would expect?

There is a strong positive linear correlation between the spreading rate and plume incidence, and therefore between spreading rate and hydrothermal venting. This makes sense, because the faster the spreading rate, the greater the rate of generation of oceanic crust and therefore the greater the amount of heat available to drive hydrothermal circulation.*

5.4.3 EVENT PLUMES

The plumes discussed so far are present more or less continuously, and so represent the steady-state venting of effluent, at least on decadal timescales. However, from time to time much larger but transient plumes have been observed. These are described as **event plumes** or, sometimes, megaplumes. Most are probably a consequence of sea-floor eruptions of lava, and several

* However, recent studies in the Indian Ocean, where sea-floor spreading rates are among the lowest in the world, have revealed the presence of several hydrothermal vent fields. This seems to suggest that there is in fact *no* significant correlation between spreading rate and the incidence of hydrothermal activity at ocean ridge axes.

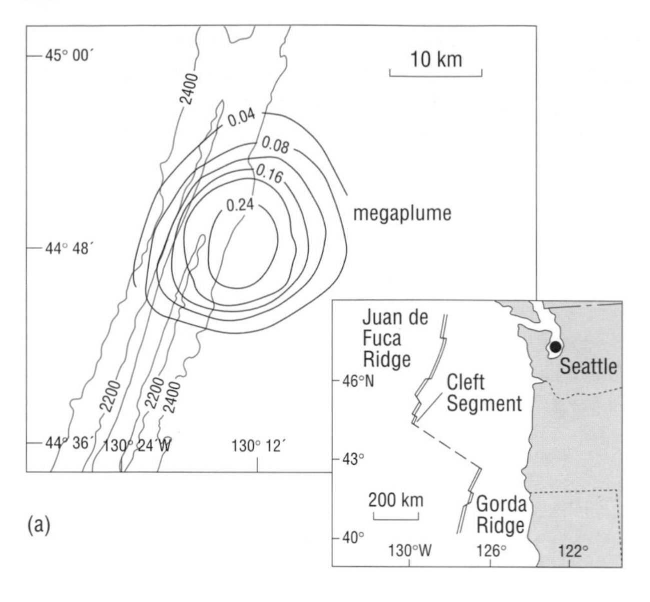

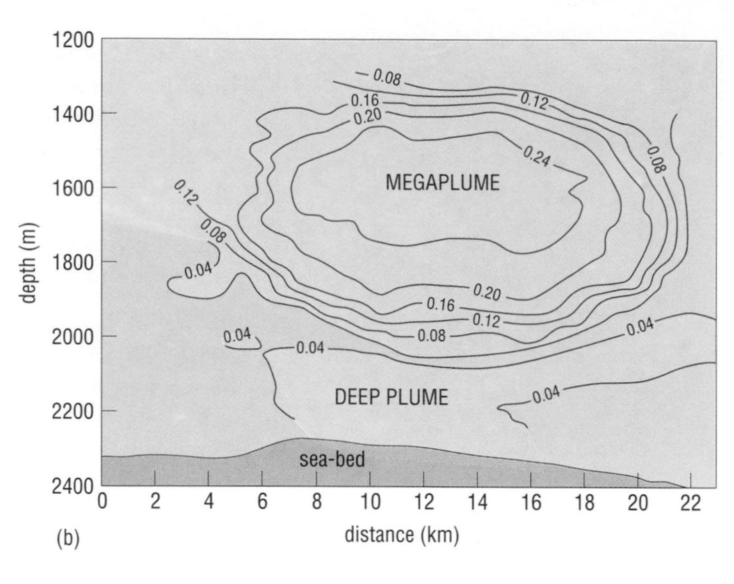

Figure 5.15 An event plume in 1986 on the Cleft Segment of the Juan de Fuca Ridge (see inset location map). Contours shown are temperature anomaly (°C) relative to ambient water at the same depth. (a) Plan view, showing also the 2400 m and 2200 m bathymetric contours; (b) north–south cross-section, which reveals the smaller, steady-state plume ('deep plume') below the event plume ('megaplume').

(like the one shown in Figure 5.15) have indeed been shown to coincide with formation of new volcanic mounds or lava flows (Figure 5.16). Whereas a steady-state plume rises only a few hundred metres, an event plume often reaches a thousand metres or more above its source.

Figure 5.16 Deep-tow video picture showing the contact between newly erupted pillow lavas (dark) and older pillow lava (pale) near 42° 40′ N on the Gorda Ridge (southern part of the Juan de Fuca Ridge) in 1996. The new lava flow is 3.5 km long but only 100–200 m wide (probably fairly typical in size) and is sited along a fissure directly below an event plume revealed by anomalies in ^{3}He, temperature, and suspended particle concentrations in March 1996. The field of view is a few metres across.

5.4.4 OFF-AXIS HYDROTHERMAL CIRCULATION

Newly formed crust at a ridge axis must lie in the potential upflow zone of the axial convective system. It moves away at a few centimetres per year.

If the (half-) spreading rate averages 2 cm yr^{-1}, how far will the newly formed crust have moved from the ridge axis in 100 000 years?

It will be about 2 km away from the axis, and so will normally be well within the *downflow* zone of the convective system where it will be penetrated by cold unreacted seawater instead of hot reacted seawater. We have seen that, as in the TAG area, faulting can sometimes channel upflow into sites 2 km or so from the axis, but it is inevitable that within a few hundred thousand years at most, newly formed crust will have moved sideways into the downflow zone.

However, as shown in Figure 5.17, the story does not end there. A few kilometres off-axis, convection cells appear to become 'locked into' the rock through which they circulate. Upflow and downflow zones remain fixed in relation to the rock, and hydrothermal circulation continues, diminishing in intensity as the crust cools with age and is moved further from the spreading axis. Much less is known about convection away from ridge axes, but detailed measurements of heat flow provide clear evidence that it occurs.

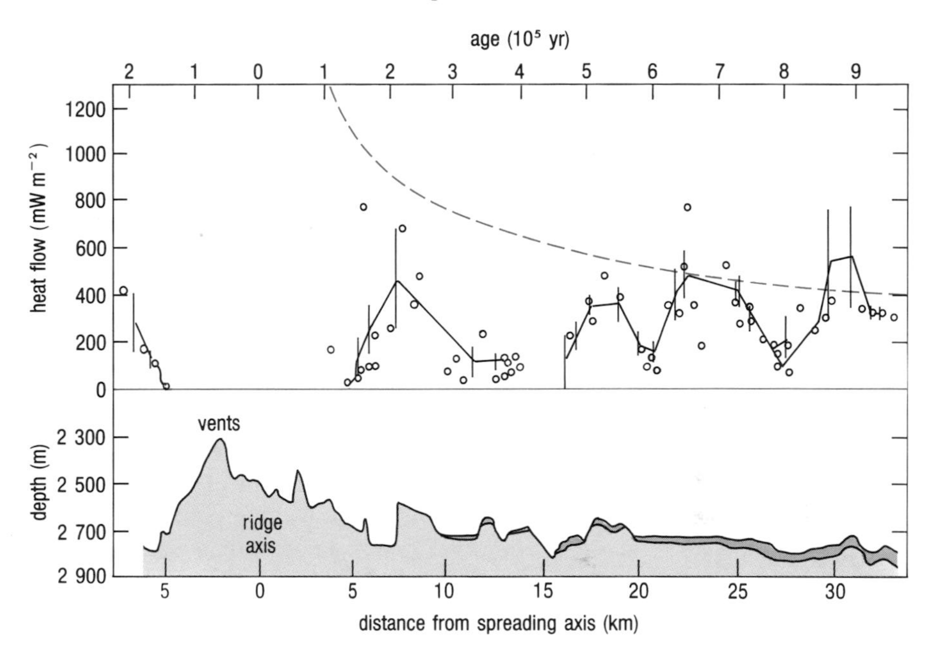

Figure 5.17 The presence of convection cells away from ridge axes can be mapped using heat-flow measurements (see Question 5.6). Open circles are individual measurements of the heat flow by conduction, and vertical bars show the range of several measurements. The broken line is the predicted heat flow, assuming conduction alone. The lower profile shows topography with progressively thickening sediment cover. These data are from the Galápagos spreading axis in the eastern equatorial Pacific. Note that the age scale is in 10^5 years, so this represents a very enlarged version of the left-hand edge of the graphs in Figure 5.6.

In some places, there is evidence that off-axis circulation permeates both igneous rocks and sediments, though little if any water is exchanged with the overlying seawater. Elsewhere, it seems that it may be continuing beneath a thick insulating cover of sediments, cut off from contact with the overlying seawater: the water simply goes round and round without being replenished until the system cools down enough to stop.

QUESTION 5.6 Think carefully about what the heat flow measurements in Figure 5.17 represent. Do the peaks or the troughs in the plotted heat flow data represent the upflow or downflow zones of convective cells?

At some drill sites, when the sedimentary seal above the isolated but still-active circulation system is punctured by drilling, bottom seawater is actually sucked down into the hole. This must be because the circulation systems are no longer in equilibrium with the prevailing hydrostatic pressure field but are underpressured relative to their surroundings.

Plots similar to that in Figure 5.17 have been obtained in oceanic crust as old as 55 Ma, which is consistent with the extent of hydrothermal circulation inferred from Figure 5.6. The phenomenon is therefore not confined exclusively to the main ridge systems and their flanking regions.

Fully one-third of the world's oceanic crust contains active circulation systems which are the descendants of hydrothermal systems that originated at ridge crests. With such a long history of hydrothermal activity affecting the oceanic crust, it is not surprising that the chemical changes in both rocks and seawater are extremely complex in detail and can vary a good deal from place to place (e.g. Sections 5.2.1 and 5.3.3).

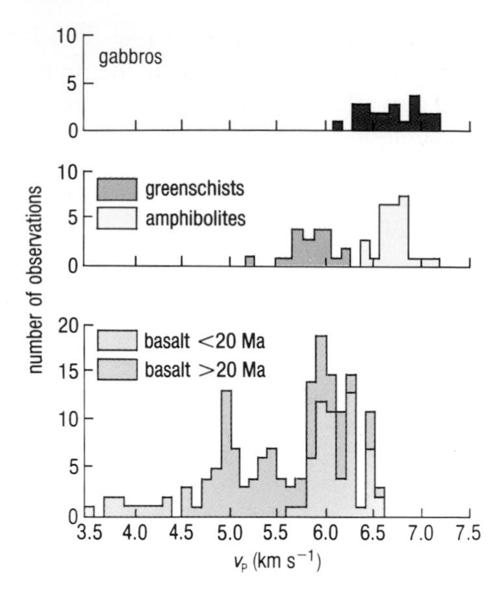

Figure 5.18 Laboratory measurements of compressional wave (P-wave) velocities at 1 kbar for water-saturated oceanic rocks.

5.4.5 THE EXTENT OF HYDROTHERMAL METAMORPHISM

We have seen evidence of hydrothermal circulation in many parts of the oceans, so just what is the extent of hydrothermal metamorphism within the oceanic crust? Figure 5.7 implies that temperatures in the top 500 m or so are generally too low for metamorphism to occur except in the immediate vicinity of black smoker conduits (remember from Section 5.2.1 that metamorphism to greenschist grade requires temperatures of 200–400 °C). However, temperatures are well within the greenschist range throughout at least the lower half of seismic layer 2, so we might expect the rock there to have been extensively metamorphosed. Unfortunately, seismic methods cannot confirm this.

QUESTION 5.7 Look at Figure 5.18 and explain why seismic methods cannot tell you whether oceanic crust has been metamorphosed or not.

So, to prove the extent of metamorphism in the oceanic crust, we need samples of the rock itself. Samples collected by dredging and submersibles are usually from the uppermost crust, subject only to sea-floor weathering, although rocks may be brought up to the sea-bed from greater depths in the crust by faulting. Both metamorphosed and unmetamorphosed rocks have been recovered from regions of faulting, but they cannot necessarily be regarded as typical, because of the possibility that large faults act to concentrate the hydrothermal circulation locally. It is rare for boreholes to penetrate more than a few hundred metres into the igneous oceanic crust, but the 1.8-km-deep ODP Hole 504B, which you met at the start of Chapter 4, is a notable exception. This initially went through pillow lavas and then dykes metamorphosed to greenschist grade, and then dykes metamorphosed to amphibolite grade in the lower part of the hole. In ophiolite complexes, it is usual for metamorphism to increase downwards from none (sea-floor weathering only) in the upper part of the pillow lavas through zeolite-bearing rocks to greenschist grade and sometimes amphibolite grade in the sheeted dykes and gabbros. This strongly suggests that the type of hydrothermal metamorphism we have been discussing is prevalent throughout much of the oceanic crust.

5.5 MASS TRANSFER BY HYDROTHERMAL CIRCULATION

Figure 5.6 shows that heat is lost mostly from spreading axes and that the convective 'deficit' itself decreases exponentially away from them. The total heat deficit, i.e. the total heat lost by hydrothermal circulation in the oceans (the 'convective heat-transfer rate'), is estimated by various authorities to be between about 8×10^{19} J and 3×10^{20} J per year. It can be shown that a volume of seawater equivalent to the whole oceans must circulate through the oceanic crust in a period of the order of 10 million years to account for this.

The rate of flow of water through the crust, F (in kg yr^{-1}), necessary to achieve the required heat transfer is given by the formula:

$$F = \frac{H}{c_w\,(T_2 - T_1)} \tag{5.1}$$

where H is the convective heat-transfer rate given above; c_w is the specific heat of seawater: $4.2 \times 10^3\,\mathrm{J\,kg^{-1}\,°C^{-1}}$; T_1 is the initial temperature of seawater at the ocean floor, e.g. 2 °C; and T_2 is the temperature of seawater after circulation through the crust, e.g. 300 °C.

We have suggested a figure of 300 °C for T_2 because (as we have seen) most of the heat is lost from the narrow zone on either side of the ridge axis. Highest temperatures are reached by the black smokers on the axial rift region, but temperatures will be substantially lower on the flanks.

QUESTION 5.8

(a) With this information, and taking an approximate average value of $2 \times 10^{20}\,\mathrm{J\,yr^{-1}}$ for H, use Equation 5.1 to work out the mass of seawater flowing through oceanic crust in the vicinity of ocean ridges each year.

(b) There is about 1.4×10^{21} kg of water in the oceans. Show that, according to your answer to part (a), this amount of seawater could be cycled through oceanic crust in about 10 million years.

The value you calculated for annual seawater flux through the oceanic crust is probably as good as any other, but the calculations were for crust close to spreading axes. It is very difficult to estimate the rate of low-temperature circulation through areas of crust that are *not* close to spreading axes. For example, off-axis circulation (Section 5.4.4) is unlikely to be *completely* sealed in, but the rate at which seawater leaks into and out of off-axis circulation systems is very poorly known. For comparison, one published estimate gives $9 \times 10^{13}\,\mathrm{kg\,yr^{-1}}$ for the near-axis seawater flux (only a little less than the value you should have calculated in Question 5.8(a)) and $2.5 \times 10^{15}\,\mathrm{kg\,yr^{-1}}$ for the off-axis flux.

Assuming your rate of 1.6×10^{14} kg of seawater circulating through near-axis oceanic crust each year is correct, and that all of this acquires the concentration of Ca^{2+} indicated in Table 5.2(b), then an extra 460 p.p.m. of calcium goes into solution. Thus, a total of about 7.4×10^{10} kg of calcium ($460 \times 10^{-6} \times 1.6 \times 10^{14}$) could be added to the oceans annually in this way. That is not very much less than the 5×10^{11} kg of calcium introduced from rivers. However, some of the hydrothermal calcium is reprecipitated in oceanic crust, as anhydrite ($CaSO_4$) and as calcite (calcium carbonate, $CaCO_3$) (Sections 5.2.2, 5.3.1). Rather more sophisticated calculations of element fluxes have shown that hydrothermal activity is the major source of lithium, rubidium and manganese to the oceans, and is an important contributor of barium and silicon, as well as of calcium. It also provides the major sink (removal mechanism) for Mg^{2+} and SO_4^{2-} introduced into the oceans by rivers.

Why have we not included potassium in this catalogue, as it is clearly important, according to Table 5.2(b)?

In Section 5.2.2, you read that potassium is leached from rocks by high-temperature seawater, but is added to rocks at temperatures below about 150 °C. Therefore, it is particularly difficult to assess the net flux of potassium between oceanic crust and seawater.

In the next Chapter, you will see how information from sediment sequences can be used to reveal changes in oceanic circulation and sea-level, in response to changes in basin shape (Chapter 3) and climatic variations.

5.6 SUMMARY OF CHAPTER 5

1 Direct evidence of hydrothermal circulation through oceanic crust was not obtained until the late 1970s, but it had been predicted from several lines of indirect evidence for at least a decade before. These included the distribution of metal-rich sediments at ridge crests, metamorphic rocks dredged from ridge crests, a major conductive heat flow deficit along the whole ocean-ridge system and laboratory experiments on basalt–seawater interactions at high temperatures and sea-floor pressure.

2 All hydrothermal systems require a permeable layer of rock which allows cold water to percolate slowly downwards over a wide area, a localized heat source below the permeable layer, and channelways for the heated buoyant (low-density) plume of hot water to escape more rapidly at the surface above the heat source.

3 Rock metamorphism during hydrothermal activity involves hydration, the addition of large amounts of magnesium, the loss of calcium and sometimes sodium and potassium, and also gains and losses of numerous minor and trace elements. The normal basaltic mineral assemblage (plagioclase, pyroxene ± olivine, and basaltic glass) is transformed into various combinations of albite, chlorite, actinolite, zeolite, quartz and other minerals, depending on the temperature and pressure conditions. Serpentinite is formed when hot seawater penetrates to seismic layer 4 and hydrates the peridotite. It behaves plastically and is of low density, so it can be forced upwards along fractures by pressure. Eventually, it may be exposed at the sea-bed. Metamorphic rocks formed in oceanic crust may later be exposed along scarps of faults and fractures.

4 During hydrothermal circulation, seawater loses all its magnesium and sulphate (which is reduced to sulphide), and gains significant amounts of calcium and sometimes also potassium and sodium, as well as several minor elements, especially silicon, but also others (e.g. barium, rubidium, iron and manganese). When hydrothermal solutions mix with normal seawater at vents, some of these constituents are precipitated, in cracks and fissures within the rock, around vents to build chimneys, and in the form of particles forming the 'smoke' at black smokers and white smokers. Sulphide is precipitated as metal sulphides (chiefly of iron, but also notably of copper and zinc), calcium as calcium sulphate (anhydrite) and calcium carbonate (calcite), barium is precipitated as barium sulphate (barite), and silicon is precipitated as silica (including quartz). Manganese is precipitated as an oxide, as is some of the iron, though most of this goes into sulphides.

5 Sea-floor weathering occurs at bottom-water temperatures and involves hydration, alteration of feldspars and glass to clay minerals and (sometimes) zeolites, as well as oxidation, especially of iron and manganese to form oxide and hydroxide coatings.

6 Hydrothermal solutions are more acid and reducing than seawater, and carry sulphide ions (see item 4 above). There is a continuum of types of hydrothermal vent: black smokers, emerging at temperatures of about 350 °C or above and precipitating mainly sulphide particles, which build vent chimneys and form black 'smoke' plumes; white smokers, with exit temperatures of about 30–330 °C and precipitating mainly sulphate particles; and warm-water vents, with exit temperatures only a few degrees above normal bottom-water temperature. The latter two types are mixtures of normal seawater (which saturates the permeable upper crust), with the high-temperature solutions that elsewhere form black smokers.

7 Hydrothermal convection occurs throughout the ocean-ridge system and extends out to crust up to about 70 Ma old, with exponentially decreasing intensity away from ridge crests. Heat-loss calculations show that the equivalent of the whole ocean volume is cycled through oceanic crust in the space of ten million years or less, and the resulting fluxes of some elements into and out of the crust are comparable with (or may exceed) those of river transport to the oceans.

8 ^3He is a ubiquitous constituent of hydrothermal fluids, being released from the primordial 'store' within the Earth when the upper mantle partially melts below spreading axes. It is an ideal tracer, for it has no other natural sources in the oceans; it can be used to trace the movement of hydrothermal effluents and hence to help provide information about current patterns and the whereabouts of active vents. Other gases that have been found in hydrothermal plumes are methane (the most abundant dissolved gas), hydrogen, hydrogen sulphide, carbon dioxide, carbon monoxide, and nitrous oxide. They may be of abiotic (inorganic) or microbial origin.

Now try the following questions to consolidate your understanding of this Chapter.

Table 5.3 Chemical analysis of average 'weathered' ocean ridge basalt in weight %.

SiO_2	47.92
TiO_2	1.84
Al_2O_3	15.95
Fe_2O_3	4.58
FeO	6.05
MnO	0.19
MgO	6.38
CaO	10.73
Na_2O	2.91
K_2O	0.53
H_2O	2.19
	99.27%

QUESTION 5.9 Compare Table 5.3 with Table 5.1. How well does it exemplify the main changes that occur in sea-floor weathering, as outlined in Sections 5.2.1 and 5.2.2?

QUESTION 5.10 Zeolites that are formed during metamorphism of oceanic crust (as opposed to those which form by sea-floor weathering) are stable in the approximate range 100–200 °C, at the prevailing pressures. Chlorite is stable in the range 300–400 °C. With reference to Figure 5.7, can you explain why chlorite predominates over zeolites in metamorphic rocks of the oceanic crust?

QUESTION 5.11 The data in Table 5.2 suggest that nearly 1300 p.p.m. of magnesium is lost from seawater to rock.

(a) Assuming an average flux of seawater at spreading axes of 1×10^{14} kg annually, how much magnesium is lost from the seawater each year?

(b) Annual river flux of magnesium into the oceans is 1.3×10^{11} kg yr^{-1}. What proportion of this does your answer to (a) amount to, and does it support the statement that hydrothermal circulation provides a major sink for magnesium in the oceans?

QUESTION 5.12 Which of the following statements are true, and which are false?

(a) Sea-floor weathering can only occur where hydrothermal circulation is in progress.

(b) Black smokers are most active along ridges in low latitudes because the Earth's surface temperatures are highest in equatorial regions.

(c) The lowest layers of sediment sequences deposited on oceanic crust should be the most enriched in iron and manganese.

(d) Water within the oceanic crust near ridge crests is cold bottom water saturating all the available spaces throughout crustal layer 2.

(e) At any one time, the $^3He:^4He$ ratio in seawater will be greater than that in the atmosphere.

CHAPTER 6 PALAEOCEANOGRAPHY AND SEA-LEVEL CHANGES

In terms of geological time, individual ocean basins are short-lived features of our planet, continuously changing in shape and size. The systems of currents within them are even more ephemeral, so that those of our present-day oceans are comparative newcomers to the global scene. For instance, there can have been no Gulf Stream before about 100 Ma ago, because the North Atlantic had not yet opened wide enough to accommodate the circulation system of which it is a part (Figure 3.1).

Information about palaeoceanography (the history of the world's oceans) comes from two main sources: magnetic anomalies and related data, from which we can deduce the changing shape of the basins themselves; and the sediments of the sea-bed, which retain a record of past events in the overlying waters. The sedimentary record also helps to unravel the history of sea-level changes on a variety of time-scales. This is a matter of considerable concern for nations with low-lying coastlines, and is of critical importance for hydrocarbon exploration on continental margins.

6.1 THE DISTRIBUTION OF SEDIMENTS

The thickness of sediments forming seismic layer 1 of the oceanic crust increases with distance from spreading axes.

What is the obvious reason for this?

The further the crust is from the ridge, the older it is and the more time has elapsed for sediments to accumulate. Near spreading axes, sediments are no more than a metre or two thick, even in depressions in the rugged topography, except where an axis lies close to land (for example in the Gulf of California). By contrast, in abyssal plain areas sediment thicknesses of a kilometre or more are commonplace (Figure 2.20). The continental shelf–slope–rise region may be blanketed by more than 10 km of sediment.

Figure 6.1 summarizes the distribution of the principal types of ocean-floor sediment being deposited at the present time. Areas occupied by shelf–slope–rise sediments are left blank, but the Figure distinguishes several types of sediment that are being deposited on the deep-sea-floor. Sediments that settle from suspension in the open oceans are known as **pelagic sediments** and, away from the polar regions, there are three main types, as indicated in Figure 6.1:

1 Calcareous biogenic (i.e. of biological origin) sediments, dominated by the remains of skeletal and shelly hard parts of **planktonic** organisms (i.e. organisms which drift passively within surface waters) made of calcite or aragonite (both forms of calcium carbonate, $CaCO_3$).

2 Siliceous biogenic sediments, dominated by the remains of skeletal hard parts of planktonic organisms formed of silica (SiO_2).

3 Red clays, dominated by clay minerals and with a relatively small proportion of biogenic material, which may be calcareous, siliceous, or both.

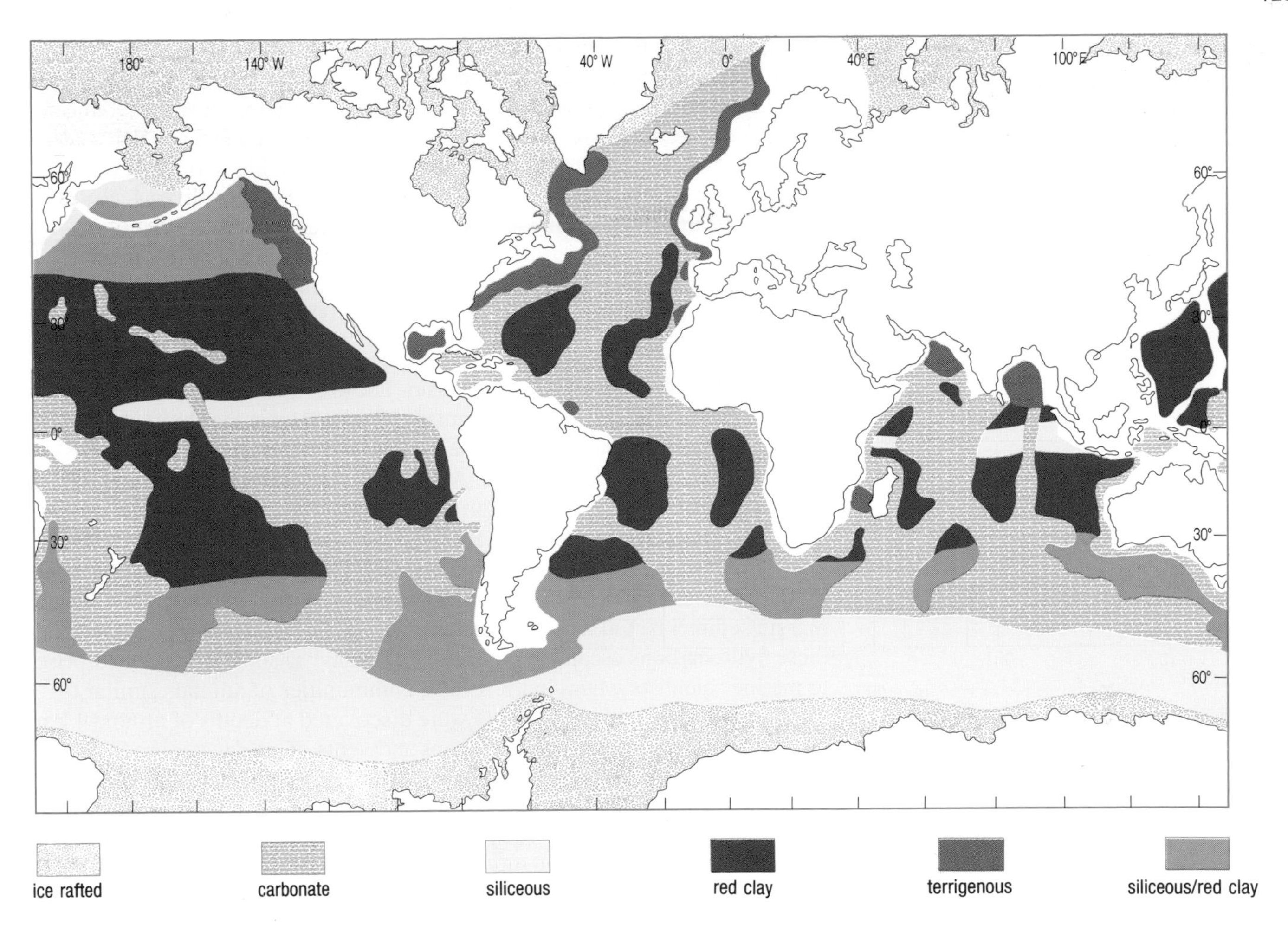

Figure 6.1 Distribution of dominant sediment types on the floor of the present-day oceans. Terrigenous sediments are those derived from land (chiefly transported into the oceans by rivers).

The red colour is caused by small amounts of ferric iron oxide. These sediments contain the smallest particles produced by continental weathering and erosion, which are carried to the deep sea mainly by winds but also by currents. They also contain the fall-out of fine volcanic ash from large eruptions (carried around the globe by winds), micrometeorites, and particles derived by the ablation of meteorites in the atmosphere. Meteorite-derived material accumulates at a rate equivalent to 0.1–1.0 mm per million years.

Sediments are described as biogenic when the rate at which biogenic components accumulate is large compared with the background 'rain' of red clay material. Calcareous biogenic sediments may contain some admixture of siliceous material and *vice versa*, and both may contain as much as 50% or more of clay as well. Biogenic pelagic sediments used to be called oozes, a term which, although it describes them rather well, is falling into disuse.

The distribution of different types of sediment is principally controlled by three interrelated factors: (i) climatic and current patterns; (ii) the distribution of nutrients and hence of organic (biological) production in surface waters; and (iii) the relative solubilities of calcium carbonate and silica, as skeletal remains sink from the surface to the sea-bed. The dissolution of calcium carbonate in seawater is strongly depth-dependent (i.e. pressure-dependent) and, contrary to what you might expect, calcium carbonate is more soluble in cold than in warm water. Both these factors account for the absence of calcium carbonate from sediments in deeper parts of the oceans, and its relative abundance along ocean ridges.

The sediments of the deep-sea-floor are deposited on the igneous oceanic crust. Their lowest layers almost everywhere consist of metalliferous sediments, deposited from black smokers and white smokers when the crust was young and still near a spreading axis. You know from Chapter 5, Section 5.4.4, that as deep-sea sediments increase in thickness with distance from ridges axes, they help to seal hydrothermal systems from further interaction with the overlying seawater. The lower parts of the sediment sequences can be affected by the warm solutions, resulting in some alteration and recrystallization of the sediments. However, as you also read in Chapter 5, hydrothermal circulation has ceased by the time oceanic crust reaches an age of around 70 Ma.

In Section 3.2, we saw that evaporite salt deposits may form during the early stages of development of an ocean basin. These evaporites are subsequently buried by the thick sediment accumulations which form the shelf–slope–rise region. The salts are not only less dense than the sediments and but also deform plastically under pressure. As a result, they can be forced upwards, doming and then punching through the sediments (Figure 6.2). These salt pillars or domes are fruitful sites for hydrocarbon exploration, as they often provide seals for oil and gas accumulations. The oil and gas are hydrocarbons produced by anaerobic decay of organic remains within the thick continental margin sediments. Oil and gas seeps occur along continental margins where these hydrocarbons escape at the sea-bed. It came as something of a surprise to marine scientists when, in the 1980s, communities of animals similar to those found at hydrothermal vents were discovered at depths of around 1 km along continental margins, where seeps and vents of *cold* water are accompanied by emission of hydrogen sulphide and hydrocarbons (mainly methane). The animal communities depend principally on chemosynthetic sulphide-oxidizing bacteria similar to those in hydrothermal vent communities (methane-oxidizing bacteria may also be present). The hydrogen sulphide and hydrocarbons are produced by the anaerobic bacterial (microbial) decomposition of organic matter within sedimentary sequences, and are forced out (along with the water) by compaction of the sediment.

The geological record of rocks exposed on the continents (in particular, the sediments associated with ophiolites) shows that sediments similar to those in

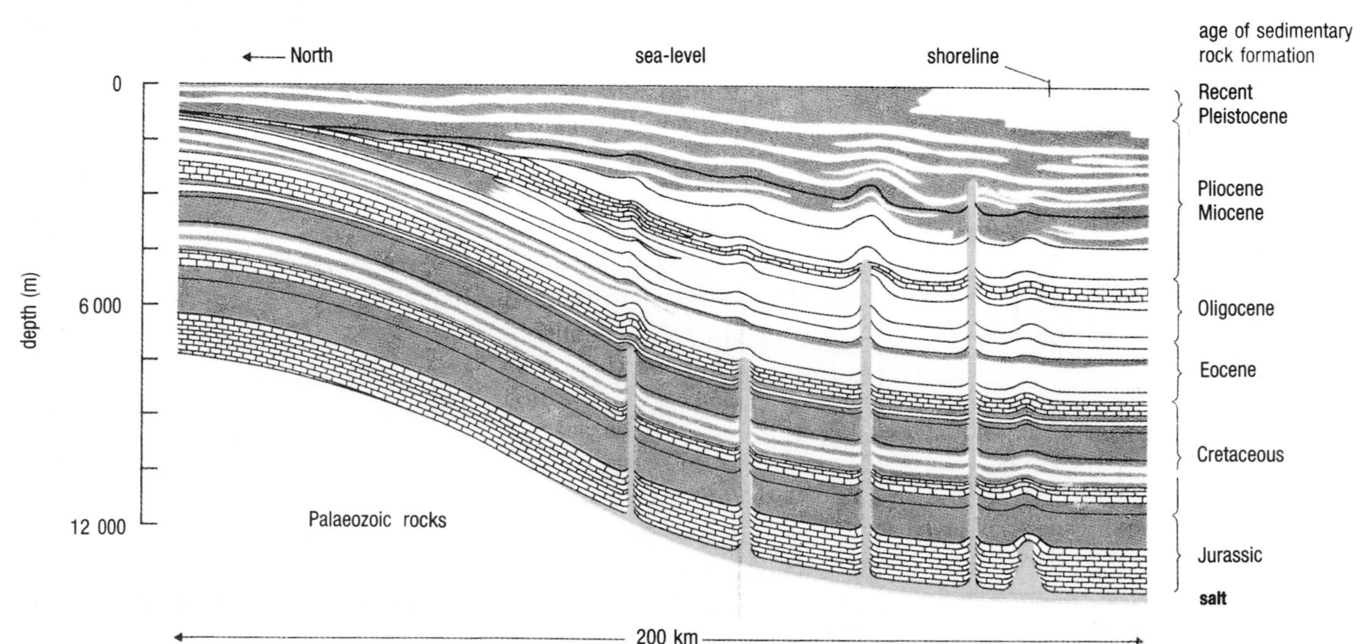

Figure 6.2 Salt domes beneath eastern Louisiana (Gulf of Mexico). Some of the domes have risen through more than 10 000 m of sediment from salt deposits at the base of the sequence. This situation is typical of many continental-shelf regions. Note the highly exaggerated vertical scale.

Figure 6.1 have been deposited on the floors of ocean basins for many hundreds of millions of years, albeit in greatly varying proportions – and of course the types of organisms making up the biogenic sediments have evolved with time.

6.1.1 SEDIMENTS AND PALAEOCEANOGRAPHY

Much information about the nature and timing of events in the evolution of ocean basins can be obtained from both the sediment sequences on the sea-floor and the remains of organisms preserved within them. By way of example, we examine a short case study in palaeoceanography: the initiation of the Antarctic Circumpolar Current.

The Antarctic Circumpolar Current is one of the major oceanic current systems. It is driven by the prevailing westerly winds of high southern latitudes (hence its alternative name of the West Wind Drift) and flows continuously eastward round Antarctica. The current extends to depths of 3000–4000 m, and owes its existence partly to the fact that there is now deep water right round Antarctica. But it was not always so: there can have been no such current until after the southern continents had all separated from Antarctica. You can see from Figure 3.1 that the southern continents broke apart at various times after about 170 Ma ago. Figure 3.6 shows that Australia and Antarctica were separated by ocean floor at least 55 Ma ago, but that oceanic crust did not begin to form between South America and Antarctica until about 20 Ma ago. However, there is evidence for an open circumpolar marine connection before then, based on the spread of marine fauna throughout the Southern Ocean. The first such evidence came from a species of the foraminiferal genus *Guembelitria*, a small planktonic organism with a maximum length of only 0.15 mm (Figure 6.3). Originally it was restricted geographically to one part of the ancient Southern Ocean, and its sudden appearance in sediments right the way round Antarctica provides the time marker for the opening of a free passage.

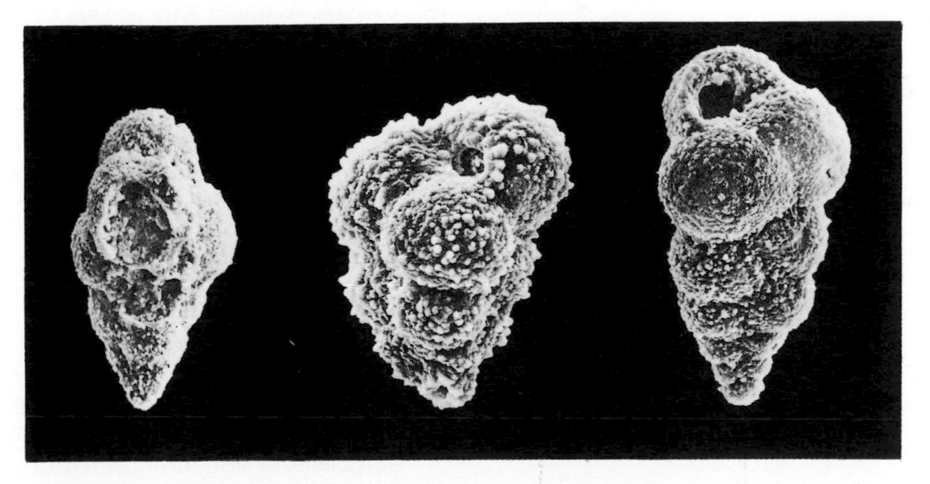

Figure 6.3 The calcium carbonate skeletal hard parts (known as 'tests') of *Guembelitria*, a planktonic foraminiferan, about 0.15 mm long. Foraminiferans are single-celled animals which may either float passively (planktonic) or live on the sea-floor (benthic).

During Lower Oligocene times (*c.* 35 Ma ago), *Guembelitria* was confined to the so-called 'Austral Gulf', between Antarctica and Australia; it has been found nowhere else in the Southern Hemisphere in sediments of Lower Oligocene age. Then in the Middle Oligocene, around 30 Ma ago, the picture changed dramatically, as Figure 6.4 shows. The distribution of *Guembelitria* suggests that a circum-Antarctic current had begun to flow, even though sea-floor spreading in the Drake Passage between Antarctica and South America did not begin until some 10 million years later.

Figure 6.4 The southern oceans about 30 Ma ago, showing the spread of *Guembelitria* (red arrows) and the initiation of the Antarctic Circumpolar Current (white arrowheads). The numbers refer to localities where *Guembelitria* is found in Middle Oligocene sediments.

It is possible, but unlikely, that *Guembelitria* spread not only eastwards into the Tasman Sea, as Australia and Antarctica separated, but also *westwards* towards southern Africa. This would not have required the Drake Passage to be open. However, there is good evidence from both marine and continental sediments that during the Middle Oligocene the world's climatic belts and wind systems occupied much the same latitude limits as they do today. As westerly winds probably prevailed near 60° S, it is unlikely that passively drifting planktonic organisms could have spread westwards against the wind-driven surface currents.

How, then, can we explain the eastward spread of *Guembelitria* round Antarctica some 10 Ma before the Drake Passage opened, according to the sea-floor age data?

Guembelitria was a planktonic organism living in surface waters. A shallow channel, perhaps only a hundred metres or so deep, would have been sufficient for it to have been transported all round the continent. Such a seaway could have been formed by rifting between South America and Antarctica (cf. Figure 3.2(b)), but before sea-floor spreading began in the Drake Passage.

Evidence from the Drake Passage itself suggests that it did not begin to open properly until about 23 Ma ago, and that deep water conditions developed there only about 18 Ma ago. Until then, the Antarctic Circumpolar Current could not have become fully established in its *present* configuration, involving flow at water depths of up to 4000 m.

We now turn our attention to another aspect of change in the ocean basins – global and regional variations of sea-level over time and their causes, which are important aspects of the evolution of ocean basins.

6.2 CHANGES IN SEA-LEVEL

It is important not to confuse the *shape* of the sea-surface, as discussed in Sections 1.2.1 and 2.6, with the *level* of the sea-surface as it is perceived and measured along coastlines. Ocean bathymetry (and its effects on the geoid) change significantly only on time-scales of 10^6–10^7 years, whereas sea-level fluctuations can occur on time-scales of decades to centuries. For many purposes, therefore, we can assume the shape of the sea-surface (the marine geoid) to be effectively constant, merely changing in overall elevation.

Sea-level is a level of equilibrium: it is simply the level to which the ocean basins between the continental blocks are filled by seawater at any particular time. In theory, it follows that if the ocean basins are interconnected then there is a world-wide mean sea-level. A change in the volume of seawater in one ocean will affect the level in all the others. Any such world-wide change in sea-level is called a **eustatic sea-level change**.

The equilibrium level is determined by:

1 The volume of water in the oceans. This in turn is controlled by:

(a) inputs from rainfall and snowfall, rivers, groundwater, melting ice and volcanism;

(b) outputs through evaporation and freezing;

(c) the temperature of the water and (to a much smaller extent) the amount of suspended matter in it.

2 The shape of the container (as reflected in the hypsographic curve, Figure 2.4). This in turn is controlled by:

(a) the global thickness and area of continental crust;

(b) the relative thermal states (and therefore densities) of the continental and oceanic crusts (especially the volume of active, and therefore hot, spreading ridges); also the volume occupied by large igneous provinces (Figure 4.25);

(c) the mass of water and sediments in the oceans and the resultant load on the oceanic crust.

QUESTION 6.1

(a) From the above lists, identify the factor that potentially has the greatest effect on the total volume of water in the oceans.

(b) Water expands in volume by a factor of $2.1 \times 10^{-4}\,°C^{-1}$. Ignoring gain and loss of water by freezing and thawing, what effect would a 1 °C increase in average ocean water temperature have on global sea-level, given that the average depth of the oceans is 3.7 km (Figure 2.4)? Is changing temperature an important factor in sea-level change?

(c) From what you know of the age–depth relationship for oceanic crust (Figure 2.13), can you explain why, other things being equal, you might expect an ocean basin with a fast-spreading axis to be on average shallower than one with a slow-spreading axis?

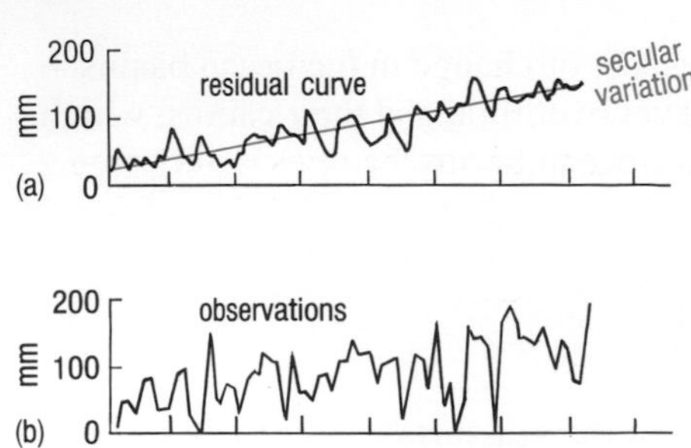

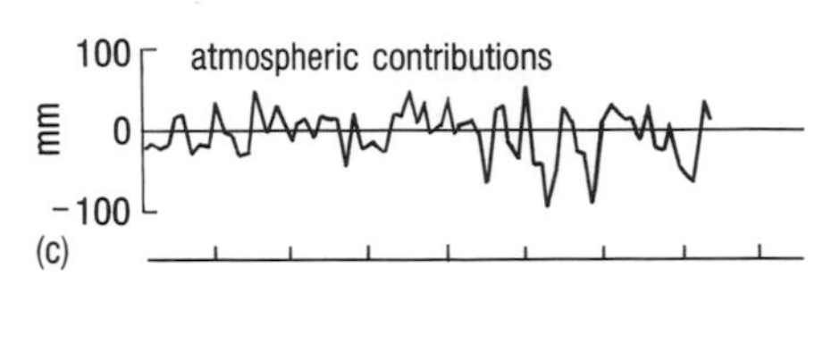

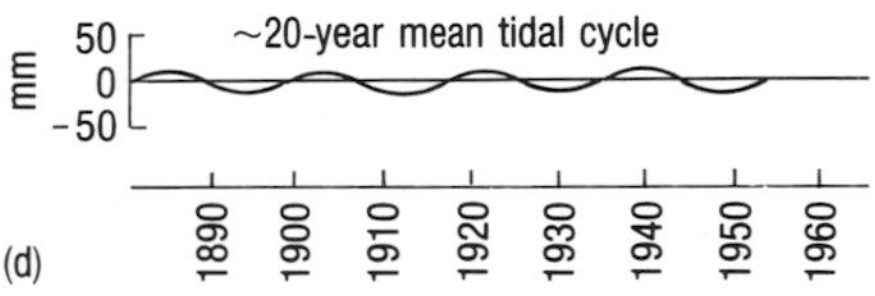

Figure 6.5 Changes in the mean sea-level at Esbjerg (Denmark), 1889–1962. (a) The progressive (secular) trend (blue line) deduced from the residual fluctuations is obtained when (b) the observed measurements are corrected for (c) changes in mean annual atmospheric pressure and (d) long-term tidal fluctuations, due to orbital and other astronomical factors. (The significance of using a graph which ends in 1962 will become apparent later.)

6.2.1 DIFFERENT TIME-SCALES IN SEA-LEVEL CHANGES

Sea-level is subject to numerous local and short-term changes, some of considerable magnitude. The principal ones are tidal fluctuations, wind-generated waves, barometrically influenced surges (see below), tsunamis (popularly, though incorrectly, called 'tidal waves'), freshwater floods, and even the waves produced by passing ships.

In spite of all these short-term variations, which may amount to fluctuations of ten or more metres in vertical range, it is still possible to define *mean* sea-level, and to measure changes in it of the order of 1 mm yr^{-1}.

For the recent past, changes in sea-level relative to a particular stretch of shore can be determined by analysis of tide gauge records. This is a complicated task because it has to take into account numerous seasonal and one-off occasional events as well as the regular tidal fluctuations, before a reliable estimate of sea-level change can be made. Figure 6.5 shows some of the principal variables that must be considered in arriving at long-term analysis of sea-level change.

The fluctuations labelled 'atmospheric contributions' in Figure 6.5(c) are of the type which, on a shorter time-scale, give rise to the barometrically influenced surges mentioned above: the lower the atmospheric pressure, the higher the local sea-level. A change in atmospheric pressure of 1 mbar will change sea-level by about 1 cm (10 mm). Atmospheric pressure at sea-level can vary from about 940 mbar or less in severe cyclones (depressions) to 1050 mbar or more in strong anticyclones. The disastrous North Sea floods of 1953 resulted from a storm surge caused by the effects of very strong onshore winds and very low atmospheric pressure coinciding with a very high tide. Yearly average atmospheric pressure differences of about 10 mbar are sufficient to account for the range of sea-level fluctuations attributed to atmospheric effects in Figure 6.5.

QUESTION 6.2 What was the average annual rise in sea-level at Esbjerg over the period shown in Figure 6.5?

The trend shown in Figure 6.5 is close to the global average rate of rise (during the same period) of 1–2 mm yr^{-1}. This rise is caused partly by the continued melting of glaciers and ice-caps as a result of warming of the climate after the most recent glacial period, and partly by expansion of near-surface ocean waters caused by the rise in average temperature (cf. Question 6.1(b)). Global temperature variations have been going on at various rates and at various scales throughout geological time. It is likely that the present rise is beginning to be exacerbated by global warming from the enhanced greenhouse effect of carbon dioxide and other gases released into the atmosphere by combustion of fossil fuels, and by deforestation which reduces the biosphere's capacity to remove carbon dioxide from the atmosphere. This progressive marine inundation now poses a major threat to the world's coastal populations.

It may be that human activities *other* than our (anthropogenic) contribution to the greenhouse effect are affecting the rate of sea-level rise, but there is no clear consensus on whether we are causing the rate to increase or to decrease. With population growth and the spread of urbanization and industrialization, humanity's water needs have greatly increased over the past 50 years. Enormous volumes of rain and river waters (hundreds of

km^3) are delayed on their oceanward path by diversion into reservoirs and irrigation projects. However, comparable volumes of water are extracted from underground aquifers at rates that greatly exceed the rates of recharge of the groundwater, and a significant proportion of this 'mined' water must eventually find its way to the sea. It is not known whether the extra amount stored annually in new reservoirs exceeds or is exceeded by the amount extracted annually from underground – indeed, the two effects could come close to cancelling each other out.

Conventional tide gauges are necessarily situated on coasts, although the global network includes stations on oceanic islands as well as along continental margins. Nowadays, these are supplemented by deep-sea pressure sensors deployed on the floors of the open oceans at locations far from coastlines. Such instruments can measure pressure fluctuations equivalent to a 1 mm change in elevation of the sea-surface even in water depths of 4–5 km. As with tide-gauge records, the data are corrected for the effects of waves, tides and atmospheric pressure variations before they can be used to assess changes in sea-level.

Unfortunately, the worldwide distribution of tide gauges and deep-sea pressure sensors (the *GLOSS* – *GLO*bal *S*ea *S*urface – network) is far from uniform, and large areas are poorly covered. In addition, a serious limitation of all Earth-bound instruments is that all of them are subject to possible crustal movements, which will cause local rises or falls of *relative* sea-level, and so they may be measuring local (**isostatic**) **sea-level changes**. We take up this topic again in Section 6.2.3, but first we look at how measurements from satellites can help.

6.2.2 USING SATELLITES TO MONITOR SEA-LEVEL CHANGES

You read in Section 1.2.1 that satellite-based radar altimeters have been used with great effect to measure the marine geoid and the departures from the mean geoid that correspond to bathymetric features in the deep oceans. The same sort of data can be used to map the height of the sea-surface relative to the satellite's orbit. Provided that the satellite's orbit is adequately known, by extending the observations globally and repeating them over a number of years the global rate of sea-level rise can be measured from space. This was achieved using radar altimeter data from a satellite called *TOPEX/Poseidon*, which was launched in 1992. It was tracked very precisely in its orbit, which was chosen to cover the entire globe between 66° S and 66° N every ten days.

Figure 6.6 is a *TOPEX/Poseidon* map showing the annual change in sea-level averaged over a three-year period, and serves to illustrate both the potential and some of the limitations of this sort of data. The potential is obvious, the extent of the coverage being limited only by the inclination of the satellite's orbit. A limitation emerges in that the most prominent features on the map – the red > 30 mm patches in the Pacific and Indian Oceans – are caused by short-term events that have no direct relationship with any general trend in mean sea-level change. For example, the local sea-level rises in the western Pacific and possibly also in the western Indian Ocean are attributable to part of an oscillation in the regional pattern of atmospheric and oceanic circulation called the El Niño–Southern Oscillation (ENSO). These oscillations are irregular, but last a few years (and so a complete cycle does not occur within the timescale of Figure 6.6); you can read about them elsewhere in this Series.

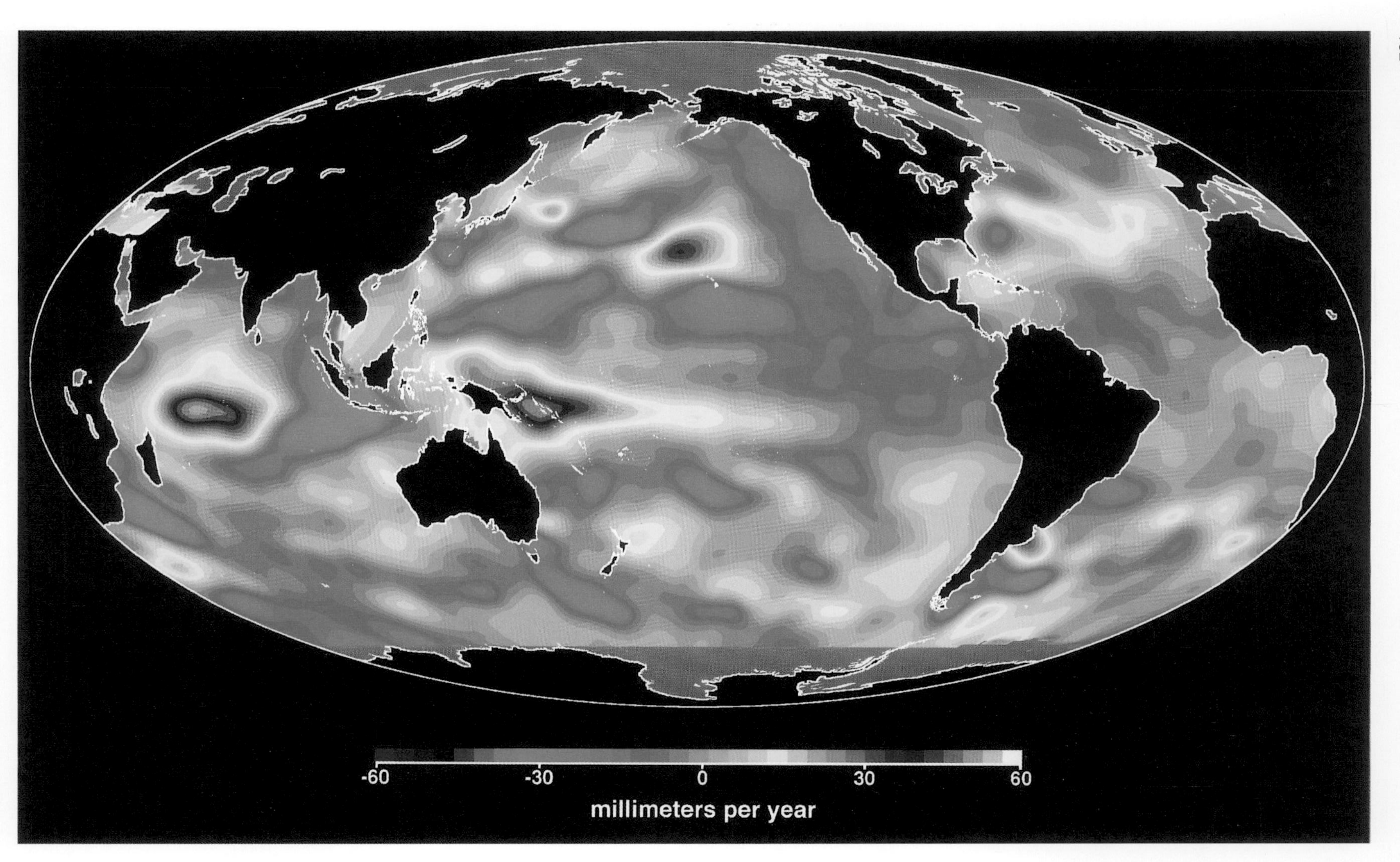

Figure 6.6 The change in sea-level derived from *TOPEX/Poseidon* radar altimeter data between September 1992 and August 1995. The average rate of change in sea-level over this period is indicated by the colour scale.

Despite the clearly anomalous areas where temporary rises and falls in sea-level (measured in tens of cm per year) can be attributed to climatic events of durations comparable with the timescale over which the *TOPEX/Poseidon* data were gathered, there remains a slight global increase in sea-level over the period covered by Figure 6.6. It is difficult to estimate this visually from such a map, although it is straightforward using the computer database from which it was compiled. Figure 6.7 and Question 6.3 provide an opportunity for you to determine the rate of global sea-level rise for yourself.

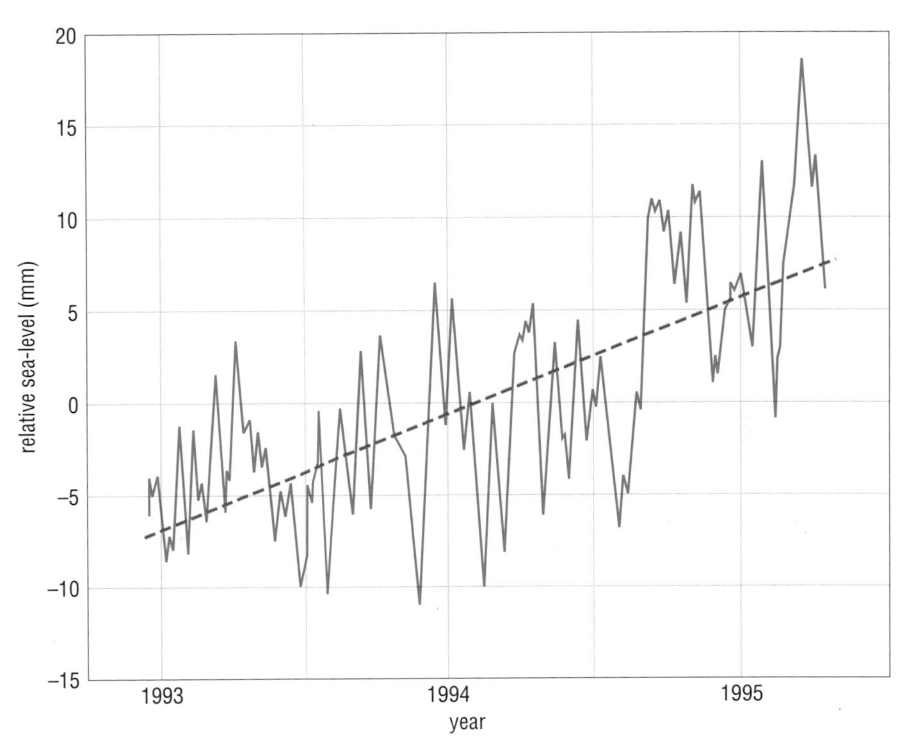

Figure 6.7 Variations in global mean sea-level derived from ten-day averages of *TOPEX/Poseidon* data. The broken blue line is the best fit to the overall trend.

QUESTION 6.3 Figure 6.7 is a plot of global sea-level derived from *TOPEX/Poseidon* data.

(a) Why do you think ten-day averages were used to draw the plot?

(b) Use the best-fit line to determine the annual rate of global sea-level rise during the period shown, according to the *TOPEX/Poseidon* data.

In fact, detailed statistical analysis of the data used to plot Figure 6.7 indicates an annual rate of sea-level rise of 5.8 ± 0.7 mm yr^{-1}. It is thought that about 2 mm yr^{-1} may be explicable by drift in the calibration of the measuring system, but there remains the inescapable conclusion that over the 1993–95 period, global sea-level was rising at something like 4 mm yr^{-1}. This rate of rise is greater than the rate you found on the basis of older data in Question 6.2, and might appear to provide confirmation of global warming, resulting in thermal expansion of seawater (and hence sea-level rise, Question 6.1(b)). However, when satellite data on global sea surface temperature are considered, it appears that this may not be the case (Figure 6.8).

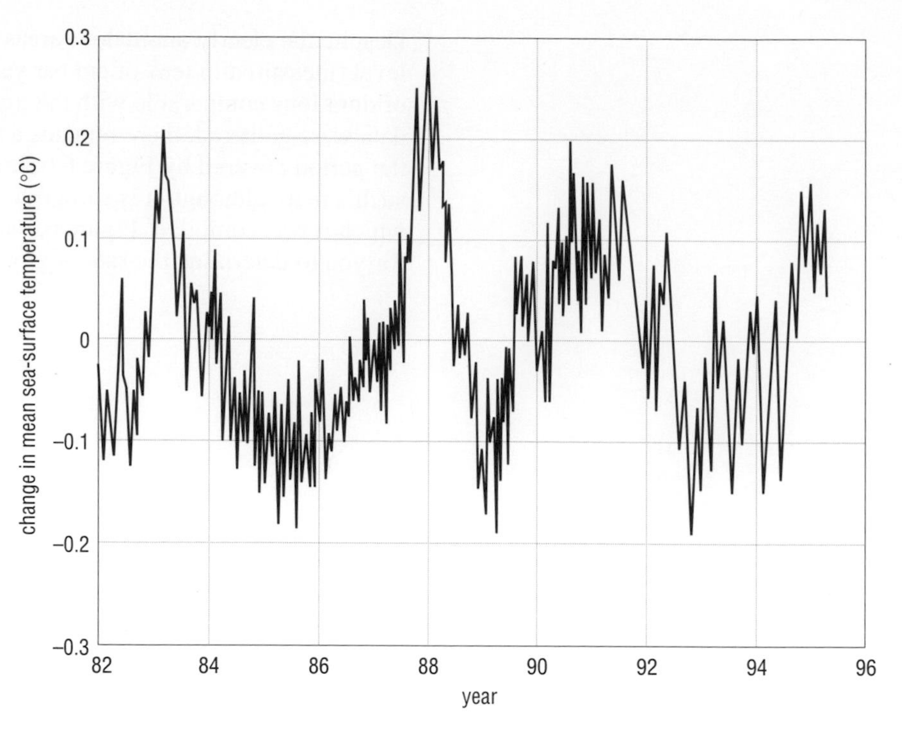

Figure 6.8 Global sea-surface temperature anomaly derived from thermal infrared mapping by satellites. There appears to be no consistent rising temperature trend over the 13-year period illustrated. The quasi-cyclic fluctuation in sea-surface temperature is a manifestation of the El Niño–Southern Oscillation (ENSO) climatic perturbation, which, although most pronounced in the tropical Pacific, is sufficiently large to show up in these globally averaged data.

Do you notice any correlation between global mean sea-surface temperature in Figure 6.8 and global mean sea-level in Figure 6.7?

Figure 6.7 covers only the latter part of the period shown in Figure 6.8. The rising trend in global sea-surface temperature during 1993 and 1994 matches the rise in mean sea-level during the same period. However, Figure 6.8 does not provide evidence for any longer-term trend in sea-level change; it only shows quasi-cyclic fluctuations in global mean sea-surface temperature (dominated by ENSO events).

It is worth considering whether the rise in global mean sea-level during 1993 and 1994 (Figure 6.7) could be a result of thermal expansion of seawater in response to the rise in global mean sea-surface temperature during the same period (Figure 6.8). Throughout the oceans, the uppermost parts of the water column are generally well mixed by winds and waves. Mixing penetrates deeper at intermediate and high latitudes than at low latitudes, but we can take 100 m to be a reasonable global average depth below which changes in sea-surface temperature can have no immediate effect. If we now do the same kind of calculation as in Question 6.1(b): assuming the upper 100 m of seawater (world-wide) warmed by 0.2 °C (estimated from Figure 6.8 between the start of 1993 and the end of 1994), we get: $100\,\text{m} \times (2.1 \times 10^{-4})\,°\text{C}^{-1} \times 0.2\,°\text{C} = 4.2 \times 10^{-3}\,\text{m} = 4.2$ mm rise in sea-level over two years, which is a rate of rise of about $2\,\text{mm}\,\text{yr}^{-1}$. This is about *half* the rate derived from the *TOPEX/Poseidon* data, as discussed in relation to Question 6.3. So, short-term thermal expansion of the uppermost 100 m of the oceans could account for about half the global sea-level rise estimated from Figure 6.8. The remaining $2\,\text{mm}\,\text{yr}^{-1}$ is the underlying long-term global average sea-level rise discussed in Section 6.2.1.

Thus, the present rate of rise in sea-level is part of a trend that has been going on for about the past 20 000 years, although the weight of evidence suggests that human activities resulting in global warming are now increasing the rate of sea-level rise. Let us turn now from the future prospects to consider past sea-level changes. In the longer perspective, sea-level has been rising and falling continuously on time-scales varying from hundreds and thousands to millions of years.

6.2.3 THE POST-GLACIAL RISE IN SEA-LEVEL

There have been several times in Earth history when ice-sheets were widespread in high latitudes. The most recent of these Ice Ages was during the Quaternary (approximately the past 1.6 Ma), and it may not be over yet. Within this period, there have been several individual **glaciations**, i.e. advances of the polar ice-caps, followed by their retreat. The most recent such glacial episode lasted from about 120 000 until about 20 000 years ago, and we now appear to be in an interglacial period (which, unless the whole Ice Age is over, we may expect to be followed by another glaciation). The effects on sea-level of post-glacial retreat of the ice have been documented in detail in many parts of the world. Note that it is only melting of *continental* ice that can lead to a rise in sea-level; melting of *floating* ice (such as that covering most of the Arctic Ocean and fringing parts of Antarctica) has no effect, because the ice is already in the sea, displacing seawater.

The initial rise in sea-level, and presumably the initial melting of the continental ice, seems to have been relatively rapid from about 18 000 years ago, gradually slowing to the current long-term rate of around 2 mm yr^{-1} about 6000 years ago (Figure 6.9). Figures 6.5 and 6.10 illustrate the modern rise for parts of the North Sea. There, only part of the continued rise is due to the increased volume of water resulting from the melting of ice; the remainder is because of the long-term subsidence of the southern North Sea basin, associated with the accumulation of 1000 m of sediment in the past million years or so. However, other local factors complicate the record of post-glacial changes in sea-level around the North Sea and make it difficult to find an independent frame of reference against which sea-level changes may be measured.

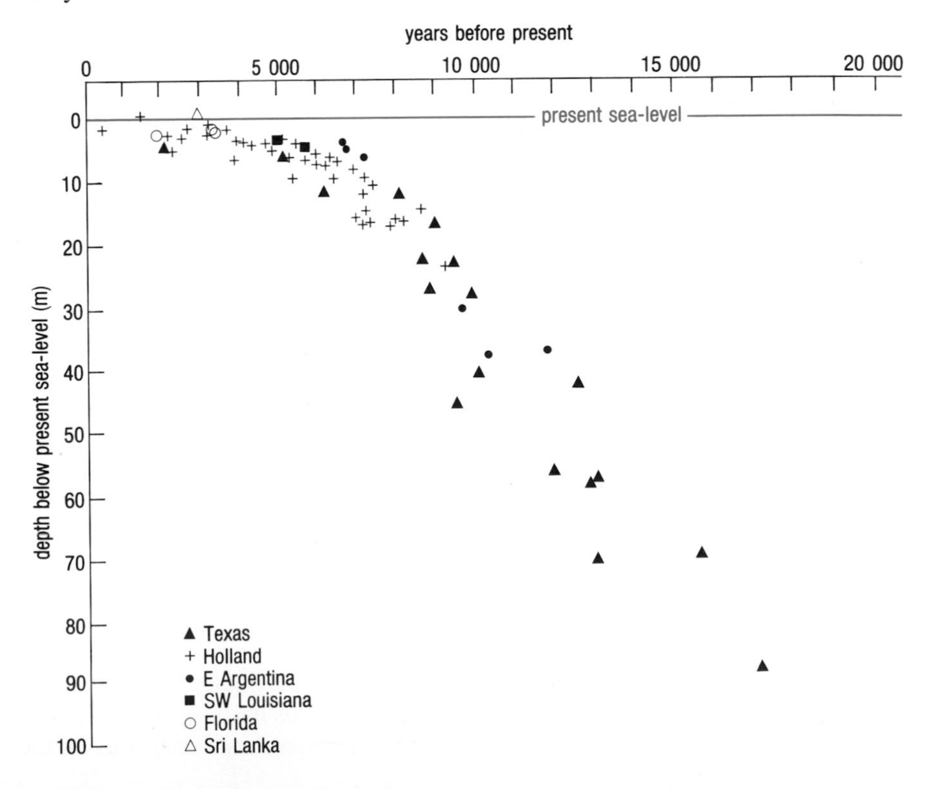

Figure 6.9 The chronology of the sea-level over the past 18 000 years deduced from ^{14}C dating of peat and the shells of shallow-water marine organisms. All locations are in stable areas of the Earth's crust, where major Earth movements have been negligible. Although there is a wide scatter in the data, the points define a *trend* for global sea-level change.

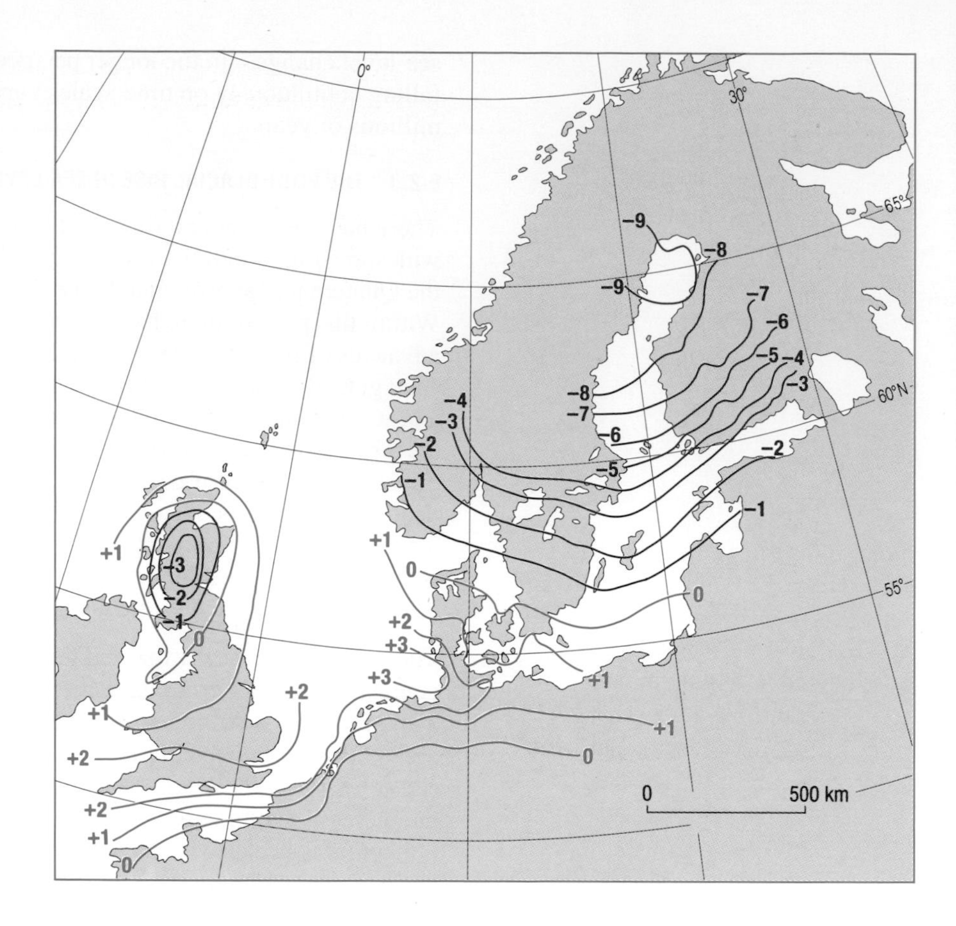

Figure 6.10 The present annual rate of change in sea-level in north-western Europe. The contours represent rises (positive) or falls (negative) in mm per year.

QUESTION 6.4 From Figure 6.10, where is sea-level still rising in north-western Europe? Can you explain why it is actually falling around most of Scandinavia?

The additional loading on the oceanic crust and continental shelves when meltwaters return to the oceans further complicates the situation at continental margins. Even in the middle of the ocean, it appears that sea-level rises more slowly (relative to an island) than would be expected from the rate of return of water to the oceans. This is because the additional weight of returning water can isostatically depress the oceanic crust.

There are, in fact, no simple answers to the question of how local isostatic adjustments combine with eustatic changes (i.e. changes in sea-level that can be identified around the world) in the record of post-glacial changes in sea-level.

6.2.4 MEASURING QUATERNARY CHANGES IN SEA-LEVEL

Data from further back in the Quaternary are insufficient to assess the contribution of isostatic adjustments to changes in sea-level. So, in a sense, the elucidation of sea-level changes becomes easier: we have only the evidence of eustatic (ocean-wide average) changes to work with. The principal feature of the Quaternary Period has been the climatic fluctuations responsible for the glaciations.

A method originally devised in the 1960s in order to estimate past marine temperatures has also proved to be of considerable value in the study of sea-level fluctuations. It depends on the differential incorporation of the ^{18}O and

^{16}O isotopes of oxygen into the calcium carbonate ($CaCO_3$) that many marine organisms use to form their shelly (skeletal) hard parts.

Oxygen isotopes and the $^{18}O : {^{16}O}$ ratio: Oxygen has three stable isotopes with atomic mass numbers of 16, 17 and 18 respectively. 99.763% of natural oxygen is made up of ^{16}O, and ^{18}O makes up most of the balance at 0.204%. Oxygen-isotope studies rely on small differences in the $^{18}O : {^{16}O}$ ratio in different samples. Although these differences are minute, they can be measured very precisely using a mass spectrometer.

We shall look at the quantitative aspects presently. First, we concentrate on the natural processes that control variations in the oxygen-isotope ratio. Marine organisms that grow skeletal hard parts of calcium carbonate incorporate different proportions of ^{18}O and ^{16}O from the water in which they live according to its temperature: the lower the temperature, the greater the $^{18}O : {^{16}O}$ ratio in the calcium carbonate that is incorporated into the skeletal material.

This ratio depends somewhat on the species of organisms, but nonetheless offers a method of measuring past temperatures of ocean waters, especially if organisms with a wide distribution can be used. Planktonic foraminiferans (e.g. Figure 6.3) are ideal, because they are abundant, widespread, and have hard parts of calcium carbonate (generally of the form known as *calcite*).

By measuring the oxygen-isotope composition of the skeletal calcite from planktonic foraminiferans found in Quaternary sediments on the ocean floor (where they accumulated upon sinking after death), it should be possible to obtain a record of fluctuations in surface-water temperatures. It should then be possible to interpret this record in terms of glacial maxima and minima, and hence in terms of eustatic changes in sea-level (Figure 6.11). However, the situation proves to be not quite as straightforward as this.

Initially, it was thought that most of the variation in the $^{18}O : {^{16}O}$ ratio for planktonic foraminiferans could be attributed simply to the temperature of the surface waters in which they lived. But when the skeletons of **benthic** (bottom-dwelling) foraminiferans in Quaternary sediments were examined, these showed almost as much variation in oxygen-isotope composition. This was a problem, because it seemed to indicate a greater variation than is likely in the temperature of oceanic bottom waters.

The temperature of present-day oceanic bottom waters nearly everywhere lies in the range of 0–2 °C. If (as seems likely) it has not changed significantly between glacial and interglacial periods within the Quaternary, how can the similarity between the variation in the oxygen-isotope ratios in calcite from both benthic and planktonic organisms be explained?

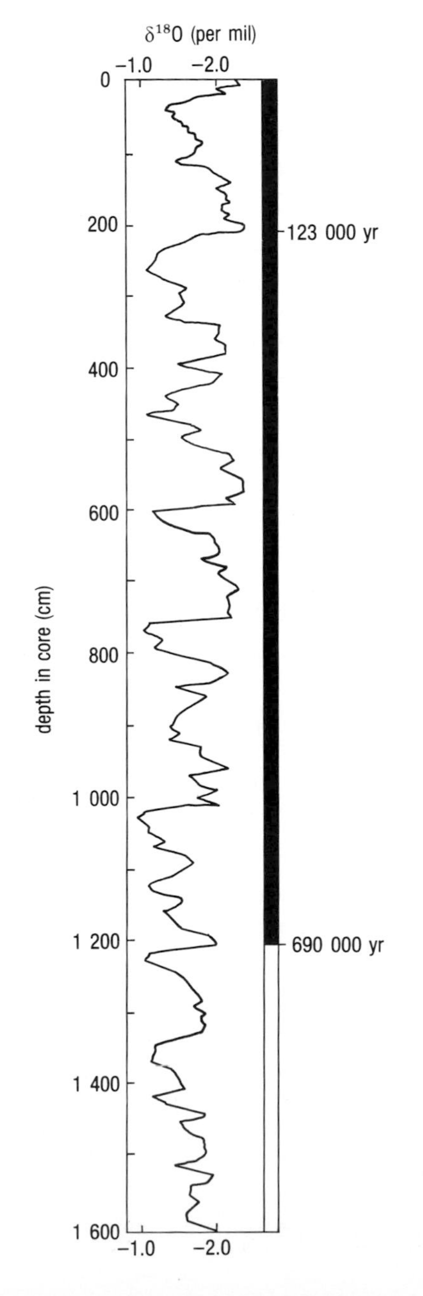

Figure 6.11 The oxygen-isotope composition (relative to a standard) of the planktonic foraminiferan *Globigerinoides sacculifera*, determined in a complete 1600 cm core spanning most of the last million years. Isotopic composition is expressed not as a simple ratio, but as $\delta^{18}O$, which is explained in the text. Plots based on benthic (bottom-dwelling) foraminiferans during the same period look almost identical. The minimum (most negative) $\delta^{18}O$ value at 123 000 years ago represents the last interglacial sea-level maximum, and the maximum (least negative) $\delta^{18}O$ value about 50 cm from the top of the core represents the last sea-level minimum before the present, during the height of the last glaciation. The black band represents the most recent interval of positive polarity in the Earth's magnetic field (the Brunhes epoch) and the white band underneath is the preceding negative polarity interval (the Matuyama epoch). Correlation with the magnetic polarity time-scale has helped in the accurate dating of the sedimentation of this and other sediment cores.

To answer that question, we begin by recalling that oxygen makes up about 90% of water by weight, and that ^{16}O is lighter than ^{18}O. Accordingly, water vapour tends to be enriched in molecules containing the lighter oxygen isotope ($H_2{}^{16}O$) and depleted in molecules with the heavier isotope ($H_2{}^{18}O$) relative to the liquid from which it evaporated (i.e. the oceans). When water vapour condenses to form rain or snow, there is a slight degree of fractionation in the opposite sense, i.e. the condensed water is slightly enriched in molecules containing the *heavier* isotope ($H_2{}^{18}O$) relative to the water remaining as vapour. However, because of the different temperatures at which the evaporation and condensation occur, the isotopic fractionation is greater during evaporation. Thus, the atmospheric evaporation–precipitation cycle causes net fractionation of oxygen isotopes, so that rainwater and snow are richer in $H_2{}^{16}O$ than the seawater from which the water vapour came.

On a global scale, when ^{16}O-enriched snow is precipitated and builds up to form glaciers and ice-caps, then the ice will be relatively depleted in ^{18}O (low $^{18}O:{}^{16}O$ ratio), while the oceans will become relatively enriched in ^{18}O (high $^{18}O:{}^{16}O$ ratio). As ice-caps grow, the proportion of ^{16}O removed from seawater (as $H_2{}^{16}O$) increases, and the $^{18}O:{}^{16}O$ ratio of the seawater must increase accordingly.

Thus, the ranges of variation in the oxygen-isotope ratio in Quaternary planktonic and benthic foraminiferans are similar because the ratios are reflecting changes in the volume of water 'locked up' in ice-caps and glaciers, not just a direct temperature effect. However, there is an implicit temperature effect in this relationship also, because the lower the global surface temperature, the more ice we would expect to be present in ice-caps and glaciers.

We must not forget the organic (biological) effect with which we started. All organisms secreting calcium carbonate have higher $^{18}O:{}^{16}O$ ratios in cold than in warm water. However, as mentioned earler, this ratio is different for different species, even in water of the same temperature (i.e. two different species in the same body of water have different $^{18}O:{}^{16}O$ ratios), but for both species the ratios are always greater in cold than in warm water.

All this may seem complicated, but the principles are simple enough, and we can summarize the main points as follows:

1 The larger the ice-caps, the greater the $^{18}O:{}^{16}O$ ratio of seawater.

2 The higher $^{18}O:{}^{16}O$ ratio in seawater will be reflected by higher ratios in the skeletons of marine organisms secreting calcium carbonate shells, and *vice versa*.

3 At any one time, although there may be differences between the $^{18}O:{}^{16}O$ ratios in the carbonate skeletons of different species living in water of identical temperature, these ratios increase in a consistent fashion if water temperature decreases (and decrease if water temperature increases).

The conventional measurement, $\delta^{18}O$: To achieve uniformity of mass spectrometer results between laboratories, the measured $^{18}O:{}^{16}O$ ratios must be calibrated against a standard sample. Historically, two standards were in common use: Standard Mean Ocean Water (SMOW) and a second one referred to as PDB. These were superseded in 1995 by Vienna Standard Mean Ocean Water (VSMOW). For our purposes, we can consider both SMOW and VSMOW as representing the average isotopic composition of normal, present-day, seawater.

Isotopic ratios are conventionally reported as δ (delta) values, which are expressed in parts per thousand (commonly referred to as 'per mil') rather than in per cent, as follows:

$$\delta^{18}O = \frac{\left(^{18}O/^{16}O\right)_{sample} - \left(^{18}O/^{16}O\right)_{standard}}{\left(^{18}O/^{16}O\right)_{standard}} \times 1000 \qquad (6.1)$$

QUESTION 6.5

(a) What would be the $\delta^{18}O$ value of a sample of Vienna Standard Mean Ocean Water itself (on the VSMOW scale)?

(b) If the standard is VSMOW, does a positive $\delta^{18}O$ value indicate enrichment or depletion in ^{18}O relative to the Vienna seawater standard?

(c) What does a negative $\delta^{18}O$ (relative to VSMOW) value tell us?

(d) Would you expect the calcareous skeletons of organisms living at the present day in cold water to have higher or lower $\delta^{18}O$ values than those of the same species living in warmer water?

(e) Would you expect the $\delta^{18}O$ value of polar ice to be positive or negative relative to VSMOW?

We can expand on your answer to part (e): the lower the temperature at which evaporation occurs, the greater the enrichment of ^{16}O in water vapour. Rain in tropical regions gives $\delta^{18}O$ values close to zero (i.e. it is isotopically very similar to VSMOW), while polar snow and ice can have values ranging from −30 per mil (the Greenland ice-cap) to −50 per mil (the South Pole).

Applications of the $\delta^{18}O$ values: To recap briefly, we can say that most of the observed variation of $\delta^{18}O$ in foraminiferal skeletons during the Quaternary must result from greater amounts of ^{16}O being incorporated into the larger volumes of ice that formed the polar ice-caps during glacial periods, leaving the oceans relatively enriched in the heavier isotope, ^{18}O. According to this interpretation, the $^{18}O : {}^{16}O$ ratio of foraminiferans, and especially that of benthic species, which live in low-temperature bottom water, can be taken as a measure of the amount of ocean water held in ice-sheets at any given time, and hence as an indicator of global sea-level.

Comparison of the oxygen-isotope composition of foraminiferans at the peak of the last glaciation (when global sea-level was at a minimum) with the composition of modern foraminiferans allows a direct relationship to be established between isotopic composition and sea-level. A difference in $\delta^{18}O$ of 0.1 per mil is found to be equivalent to a 10 m change in sea-level, and this relationship can be used to estimate global variations in sea-level resulting from changes in polar ice volumes.

QUESTION 6.6

(a) Why could the $\delta^{18}O$ values of benthic foraminiferans be considered to be more useful than those of planktonic foraminiferans in determining past changes in sea-level?

(b) Using Figure 6.11, estimate the extent of the sea-level change for (i) the last post-glacial rise in sea-level, and (ii) the preceding major rise in sea-level leading up to 123 000 years ago, and associated with the interglacial warm period.

(c) Looking at Figure 6.11, would you say that among the numerous rises and falls of sea-level in the past 800 000 years (i) the last major fluctuation was especially noteworthy and (ii) the rises are more or less abrupt than the falls?

The Quaternary fluctuations in sea-level of 100 m or more (Figure 6.11) were the result of about 50×10^6 km^3 of water being alternately withdrawn from and returned to the oceans. If the remaining continental polar ice-caps were to melt, the 30×10^6 km^3 or so of water locked up in them would raise the world's sea-level by about a further 60 m. There is good evidence that the short-term fluctuations characterizing glacial–interglacial periods are superimposed on longer-term growth and decay of the polar ice-caps. We can demonstrate this by summarizing the history of the Antarctic ice-sheet.

6.2.5 THE GROWTH OF AN ICE-SHEET: ANTARCTICA

Evidence for the rate of growth of the Antarctic ice-sheet comes mainly from the nature of the sediments on the sea-floor around Antarctica (Figure 6.1), supported by oxygen-isotope analyses of foraminiferal remains in those sediments. The distribution of ice-transported sediments around Antarctica has been well-documented. These sediments are characterized by:

1 The presence of exotic rock fragments, which can only have been brought from the Antarctic continent by ice, which then melted and deposited its sedimentary load.

2 Poor sorting (i.e. having a wide range of particle sizes), quite unlike those found along continental margins in non-glaciated areas.

3 Quartz grains with distinctive surface features that are typical of glaciated regions (and can be identified using a scanning electron microscope).

The presence of ice-transported debris in sediments of late Eocene age (*c.* 40 Ma ago) off western Antarctica suggests that this region may have been partly glaciated at that time, although the oxygen-isotope data indicate that there were no large accumulations of ice. The first definite identification of debris transported by ice is in late Oligocene sediments (*c.* 25 Ma ago) off eastern Antarctica. Since then, the distribution of glacially derived sediments has become more widespread, reaching its most extensive distribution in the Quaternary, and the proportion of glacially derived material within the sediments has increased. Thus, the Antarctic ice-sheet had already begun to develop 25 Ma ago, and probably earlier. Evidence for the growth of the ice-sheet is supported by the oxygen-isotope data from the sediments, which indicate a progressive decrease in the temperature of the Southern Ocean waters.

QUESTION 6.7

(a) According to Figure 6.12, what was the surface-water temperature when the first definite signs of ice-transported debris appeared in the sedimentary record?

(b) When would you say that the Antarctic ice-sheet began to grow rapidly towards its maximum development in the Quaternary Ice Age?

It is possible that by the late Oligocene (25 Ma ago), the Antarctic ice-sheet was already extensive enough to have had a noticeable effect on sea-level, but as you should have deduced in attempting Question 6.7, the rate of cooling increased considerably in the mid-Miocene, at which time the Antarctic ice-sheet should have begun to grow more rapidly. This

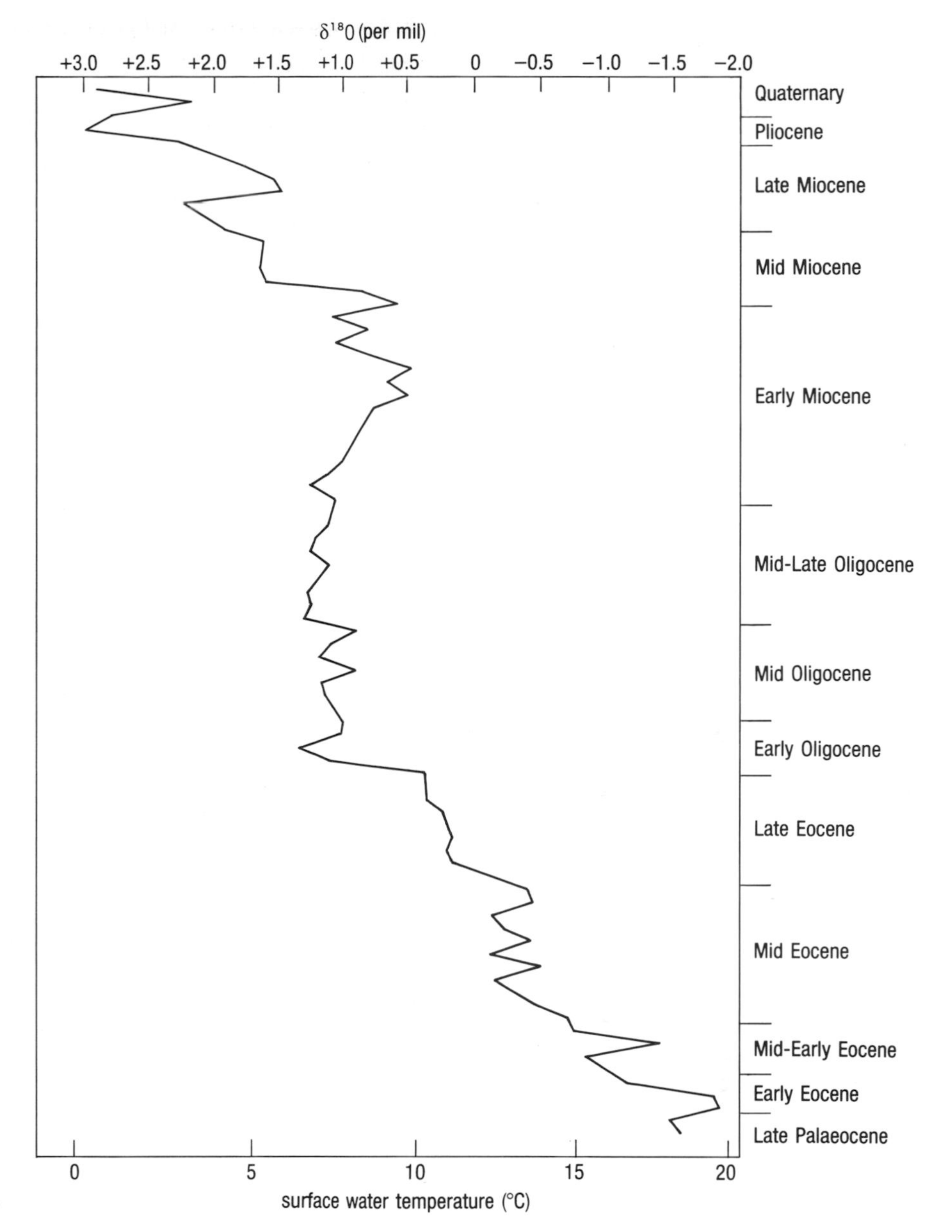

Figure 6.12 Oxygen-isotope composition and surface-water temperatures for the Antarctic region over the past 55 Ma or so, obtained from planktonic foraminiferal remains in DSDP cores. Arbitrary samples are plotted at constant intervals, so the spacing of periods along the side is not proportional to thickness of sediment or to age differences (see Appendix for geological ages in Ma). *Note*: the standard used for the isotope ratios is different from that used in Figure 6.11, which covers a much shorter time-scale and lies entirely within the Quaternary. However, the pattern would be similar whichever standard were used.

development has been linked to the final separation of South America from Antarctica, the initiation of the Antarctic Circumpolar Current (Section 6.1.1), and thermal isolation of the Antarctic continent from warmer waters to the north. In the Northern Hemisphere at about the same time, glacial conditions became more widespread in mountainous regions, although the great continental ice-sheets did not appear there until much more recently – about 3 Ma ago – by which time the Antarctic ice-sheet had already reached roughly its present size.

Even the rather coarse scale of Figure 6.12 suggests that the growth of the ice-sheet was not uniform. It shows that there were climatic fluctuations between cold and warm periods well *before* the higher frequency glacial–interglacial cycle of Figure 6.11 became established (1.5 Ma or more ago). About 5 Ma ago, such fluctuations were associated with one of the most remarkable consequences of sea-level change to be found in the geological record.

6.2.6 THE SALINITY CRISIS IN THE MEDITERRANEAN

Early in the Miocene (*c.* 20 Ma ago), the Arabian Plate impinged on the Eurasian Plate, blocking the link between the Mediterranean and the Indian Ocean to the east. The Mediterranean became an almost totally landlocked sea, with only a shallow connection to the Atlantic. This shallow strait had a tendency to close as Africa moved northwards relative to Europe. Moreover, the loss of the eastern marine connection led to the development of drier climatic conditions throughout the Mediterranean region. Evidence for this comes in part from the thick sequence of Miocene evaporites in the Red Sea basin, which was connected to the Mediterranean during most of the Miocene (Section 3.2.1).

However, towards the end of the Miocene (*c.* 5 Ma ago), during a time-period referred to as the 'Messinian', the connection between the Red Sea and the Mediterranean was broken and a passage opened up between the Red Sea and the Indian Ocean to the south. At about this time, deposition of evaporites ceased in the Red Sea but began in the Mediterranean. Great thicknesses of salts were deposited over wide areas of the Mediterranean basin during a period of about 700 000 years. This event involved the virtually complete drying up of the Mediterranean and deposition of evaporites under extremely saline conditions, and is usually referred to as the 'salinity crisis'.

It is difficult to imagine that the Mediterranean could ever have dried up completely, especially as it is now more than 3 km deep in places, with an average depth of about 1.5 km. However, this proposition is revealed as quite reasonable when we consider how long the Mediterranean might take to dry up completely, if it were to become isolated from the Atlantic Ocean again.

The surface area of the Mediterranean is about $2.5 \times 10^6\,\text{km}^2$, and its average depth of 1.5 km gives a total volume of water of about $3.75 \times 10^6\,\text{km}^3$. At these latitudes, evaporation is considerably in excess of precipitation. The present-day annual loss of water from the Mediterranean by evaporation (E) has been worked out at $4.7 \times 10^3\,\text{km}^3$, and the annual precipitation (P) is $1.2 \times 10^3\,\text{km}^3$. The annual difference between the two is:

$$\begin{aligned} E - P &= (4.7 \times 10^3\,\text{km}^3) - (1.2 \times 10^3\,\text{km}^3) \\ &= 3.5 \times 10^3\,\text{km}^3 \end{aligned}$$

About $0.25 \times 10^3\,\text{km}^3\,\text{yr}^{-1}$ of water is added to the Mediterranean by rivers and from the Black Sea, which reduces the net rate of loss to about $3.25 \times 10^3\,\text{km}^3\,\text{yr}^{-1}$. At present, of course, this net loss of water by evaporation is made good by water flowing in through the Straits of Gibraltar.

QUESTION 6.8 Suppose the Straits of Gibraltar were to be closed, cutting off this flow from the Atlantic. How long would it take for the Mediterranean to dry up?

Your answer shows that major changes affecting the Earth's surface do not always require a great time – they can sometimes happen over the span of human dynasties. There is no question that the Mediterranean *could* completely dry up in a millennium or two. The evidence that it did become dry comes mainly from buried river gorges a kilometre or so below the valleys of large present-day rivers such as the Nile and Rhône, and from the late Miocene (specifically, Messinian in age) evaporite salt deposits, thicker than 1 km, within the sediment sequences sampled by drilling over much of the Mediterranean. In many places, they form salt domes (Figure 6.13).

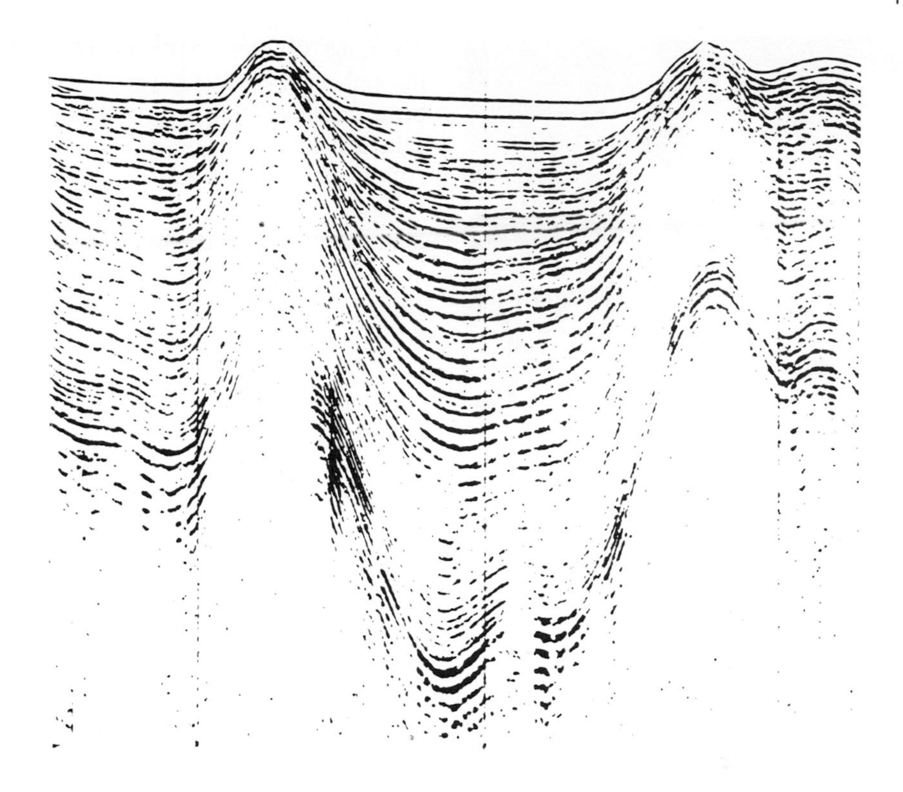

Figure 6.13 A continuous seismic reflection profile of part of the eastern Mediterranean, showing salt domes. Some protrude as knolls above the sea-bed; others are still deeply buried. The profile is about 30 km long and the vertical exaggeration is ×6 (which is approximately half the vertical exaggeration used in Figure 6.2).

We now encounter something of a problem. There is no way that the evaporation of 3.75×10^6 km^3 of seawater (the present volume of water in the basin) could have produced a 1 km-thick layer of salts on the Mediterranean floor. Every litre of normal seawater contains 35 g of dissolved salts, and there are 10^{12} litres in a cubic kilometre. So, the amount of salt in the seawater in the Mediterranean today is approximately:

$$35 \times 10^{12} \times 3.75 \times 10^6 \approx 130 \times 10^{18} \text{ g, or } 1.3 \times 10^{17} \text{ kg}$$

and it is unlikely to have been very different around the end of the Miocene.

A reasonable average density for such salts after deposition is 2×10^3 kg m^{-3}, so that amount of salts would have a volume of:

$$\frac{1.3 \times 10^{17}}{2 \times 10^3} = 6.5 \times 10^{13} \text{ m}^3 = 6.5 \times 10^4 \text{ km}^3$$

QUESTION 6.9 Assume that these evaporites accumulated over four-fifths of the Mediterranean floor, say an area of 2×10^6 km^2. What thickness would that volume of evaporites attain?

The answer to Question 6.9 suggests that the evaporation of something like 30–35 times the volume of water in the present Mediterranean would be required to produce the 1 km thickness of evaporites found in the sedimentary sequences.

Figure 6.14 The Gibraltar Waterfall. (*Artist*: Guy Billout.)

Thus, the Mediterranean cannot have been totally cut off from the Atlantic throughout the 700 000-year Messinian 'salinity crisis'. There must have been substantial influxes of Atlantic water to account for the amount of salts deposited, perhaps even enough to re-establish normal marine conditions for short periods. This is quite feasible, considering that once the Atlantic connection was broken again, the basin could dry up in little more than a thousand years. In order to enable the total thickness of evaporites to accumulate, huge volumes of seawater must have been intermittently supplied and evaporated. The basin must have been very rapidly refilled with seawater whenever the Atlantic connection was re-established; the Gibraltar 'waterfall' is thought to have been about a hundred times larger than the Victoria Falls on the Zambezi river (Figure 6.14). However, the *average* rate of supply of seawater was below the average rate of evaporation throughout the million years or so when the evaporites were being deposited.

All this changed abruptly about 4.8 Ma ago, when the supply once again began to exceed the amount removed through evaporation, and the whole basin was filled up again. This restored normal marine conditions, so that deposition reverted to muds and deep-water carbonate sediments. The Atlantic connection became deep enough for cold deep water to gain access to the Mediterranean throughout most of the Pliocene, but about a million years ago the Gibraltar sill was uplifted and this deep supply ceased. Since that time, the Mediterranean has gradually acquired its present characteristics.

Evaporites, $\delta^{18}O$ changes, ice volume and sea-level: Oxygen-isotope data have been used to relate changes in the volume of glaciers and polar ice-caps to the changing sea-levels that were responsible for bringing about the Messinian salinity crisis. The evaporites were deposited in two discrete periods. The sedimentological evidence provides abundant evidence of repeated phases of flooding, desiccation (drying out), and subaerial erosion within each period.

Towards the end of the deposition of the lower evaporites, the supply of water from the Atlantic was probably cut off completely. The resulting desiccation is represented by a widespread unconformity (an interval of non-deposition and erosion). This was in turn followed by a marine transgression (influx of the sea), when the Mediterranean may have been refilled to the brim and the deep-water sediments accumulated for a while. There was then another period of repeated desiccation when the upper evaporites were deposited. Evaporite deposition finally ceased in the early Pliocene (4.8 Ma ago), and seawater refilled the Mediterranean, restoring the conditions for pelagic (deep-water) sedimentation.

Evaporite formation had been initiated about 5.5 Ma ago when the Mediterranean became isolated from the world ocean, through the combined effects of long-term tectonic uplift in the west (associated with northward movement of the African Plate), and a global fall in sea-level. The successive inundations and desiccations during the rest of the Miocene and the earliest Pliocene (between *c.* 5.5 and 4.8 Ma ago) were controlled by fluctuating sea-levels resulting from changes in global ice volume. These changing sea-levels are revealed by oxygen-isotope measurements on samples of benthic foraminiferans from sediments with very accurately known ages, from both the Atlantic and Pacific Oceans. Some of the results and the correlation with the Messinian evaporites are shown in Figure 6.15.

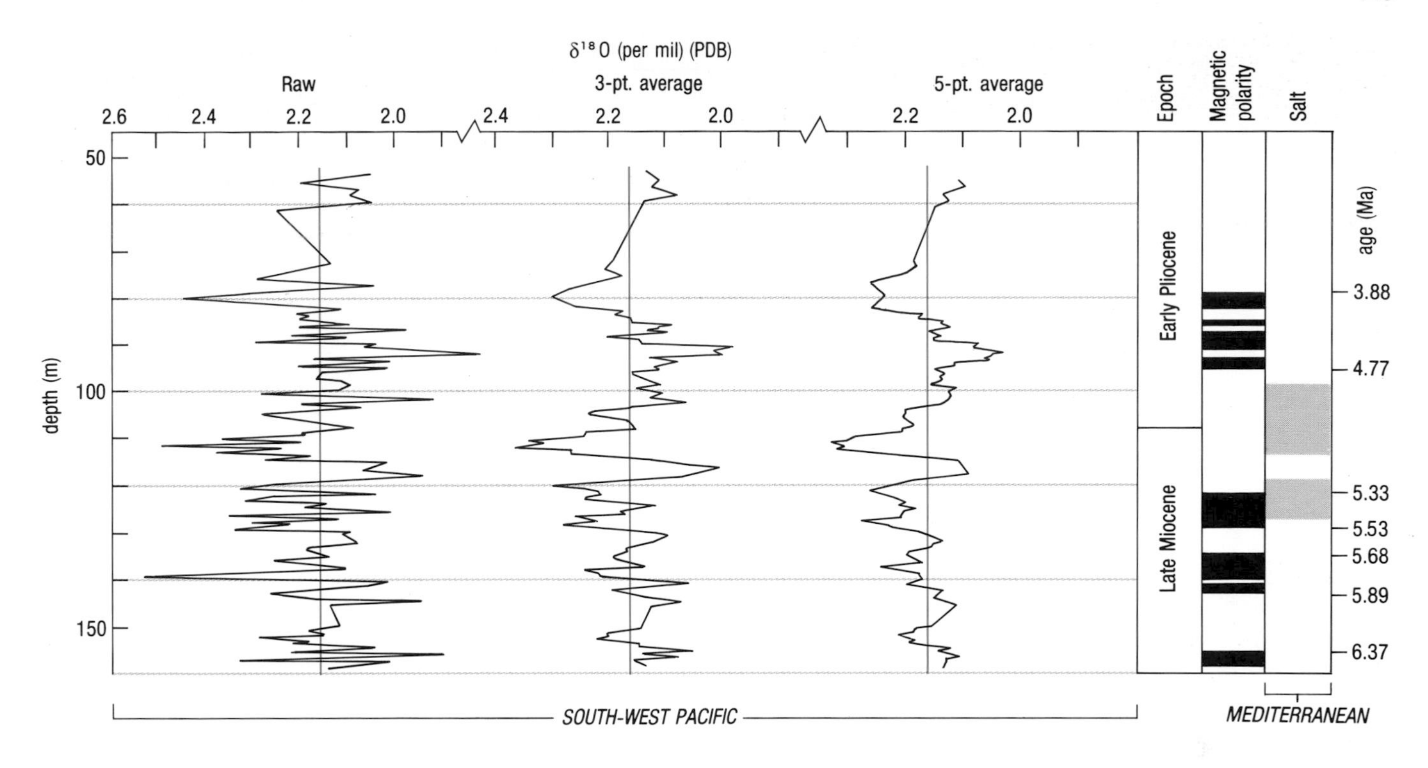

Figure 6.15 Oxygen-isotope data from the benthic foraminiferans *Planulina wuellerstorfi* and/or *Cibicdoides kullenbergi* from Site 588, DSDP Leg 90, south-west Pacific. The sampling interval between depths of 80 and 160 m is about 20 000–25 000 years. The raw $\delta^{18}O$ values clearly show that the late Miocene–early Pliocene was characterized by high-frequency variation in $\delta^{18}O$ values. The 3- and 5-point running averages of the data reveal the general trends in $\delta^{18}O$ values, and resemble glacial–interglacial cycles. The two main periods of evaporite deposition in the Mediterranean are shown for comparison.

QUESTION 6.10

(a) Looking especially at the right-hand curve of oxygen-isotope data in Figure 6.15, why do you think that $\delta^{18}O$ values were high during the two periods of evaporite deposition but low before, after and in between them?

(b) How does the left-hand curve of oxygen-isotope data help to explain the thickness of the evaporite sequences?

(c) Approximately what is the time-scale of the cycles in the left-hand curve? (Note that the age-scale on Figure 6.15 is not uniform.) How does it compare with that in Figure 6.11?

(d) What was the approximate duration of the Messinian salinity crisis, according to Figure 6.15?

(e) Figure 6.15 indicates a major fall in sea-level at around 4 Ma ago. Suggest reasons why this did not lead to yet another phase of desiccation and evaporite deposition.

(f) From your earlier reading in this Chapter, where did most of the ice probably accumulate to bring about the falls in sea-level?

Now we will consider briefly whether the uppermost part of the Miocene evaporite sequence in the Red Sea (Section 3.2.1) was deposited in the same way as the Messinian evaporites. This seems very unlikely for a variety of reasons. The Red Sea evaporites were almost certainly deposited in a shallow, subsiding basin, where sea-level was very close to what it is now throughout the Miocene. There are coral reefs along the margins of the Red Sea that provide evidence of this. The Red Sea evaporites appear to have reached practically their present thickness towards the end of the Miocene (5 Ma ago) and so they pre-date the onset of major global sea-level falls caused by the glaciations. Table 6.1 summarizes the main features of the periods of evaporite deposition in these two areas.

Table 6.1

Approximate age of event (Ma ago)	Mediterranean sea-level	Mediterranean	Red Sea
1	normal sea-level	uplift of Gibraltar sill	normal sea-level throughout
		normal marine conditions with access of cold deep Atlantic water	normal marine conditions
4.8	reduced sea-level	Gibraltar 'waterfall'; evaporite deposition (with intermittent flooding) in shallow basins far below normal sea-level	connection with Indian Ocean and isolation from Mediterranean
	normal sea-level	normal marine conditions	deposition of evaporites in shallow spreading basin connected to Mediterranean
20		isolation from Tethys Ocean to east	

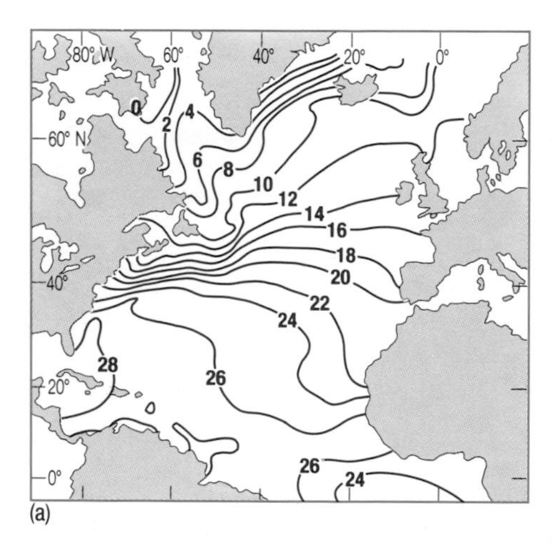

(a)

(b)

Figure 6.16 Maps comparing the distribution of surface temperatures in the northern Atlantic summer (a) at the present time and (b) 18 000 years ago. Isotherms are in °C (determined in (b) from foraminiferal plankton assemblages in about 100 deep-sea cores). The margins of sea-ice in (b) are deduced from the characteristics of the sediments and from the knowledge that the sea-level was then about 100 m lower than now.

In Section 3.3.1, we discussed the Mediterranean as an ocean in the terminal stages of its life cycle (Table 3.1, Stage 5). We must emphasize that the laying down of evaporites is not an *essential* feature of either young or old oceans. However, because young and old oceans tend to be narrow deep basins that can be isolated from the world ocean by either tectonic forces or falling sea-level (or both), deposition of evaporites is more likely to occur in them than in major ocean basins. Salt domes will occur wherever the evaporites are overlain by a sufficient thickness of other sediments (Figures 6.2 and 6.13).

6.2.7 THE MIGRATION OF CLIMATIC BELTS

You may have been wondering how warm conditions could have been sustained in the Mediterranean while ice-caps were growing at the poles. To resolve this apparent paradox, we will digress briefly to discuss climatic variations and the way in which evidence based on the distribution of oceanic sediment types and the biological remains they contain can be used to determine how climatic belts have expanded and contracted without changing their relative positions.

Different kinds of planktonic organisms can be very useful as climatic indicators (as indeed are many land plants and animals). The remains of organisms that lived in the Quaternary glacial and interglacial periods have been used to trace the expansion of cold 'high latitude' climatic zones towards the Equator and the concomitant contraction of temperate ('mid-latitude') zones as the ice-sheets advanced – and *vice versa* as they retreated. Figure 6.16 illustrates this effect, showing how present-day summer temperatures in the northern Atlantic compare with those of 18 000 years ago, at the end of the last glacial maximum.

QUESTION 6.11 Does Figure 6.16 suggest that the *tropical* climatic belt north of the Equator was significantly 'compressed', compared with its present extent, during the last glacial maximum?

It is worth noting that not all experts accept the interpretation of the study on which Figure 6.16 is based. A growing (but sometimes contradictory) body of evidence suggests that surface waters in the tropics were around 2–3 °C cooler than at present during the most recent glacial maximum (see also comment on answer to Question 6.6(a)). However, our general conclusion remains valid that it was the mid-latitude climate belts that were most compressed.

On longer time-scales, of course, evolutionary changes have occurred in the plankton, including those species used for assessing temperature fluctuations. The absence of barriers in the oceans means that planktonic organisms appear to have been able to evolve virtually synchronously around the globe. These evolutionary changes in the plankton, coupled with information from remains of other organisms and the sediments in which they occur, along with the magnetic reversal time-scale, provide powerful tools for interpreting the history of oceans. During the past 180 Ma, events about 1 Ma apart can readily be distinguished, and for the last 25 Ma this resolution can be as good as 100 000 years or better.

In Mesozoic times, the tropical climatic belts were generally wider than at present and the poles were free of ice. As implied at the end of Section 6.2.4, this means that global sea-level must have been some 60 m higher than now. The total range of sea-level fluctuations resulting from the growth and decay of ice-caps is of the order of 200 m.

The geological record preserves evidence of numerous world-wide **transgressions** and **regressions** (advances and retreats of the sea over the land), in response to rises and falls of sea-level, over periods when the climate was consistently warm and there was little or no polar ice. At the start of Section 6.2, we hinted at another cause of sea-level variation, and we now examine it in more detail.

6.2.8 THE EFFECT OF PLATE-TECTONIC PROCESSES ON SEA-LEVEL

There are two major types of plate-tectonic process that may be important in causing rises in sea-level that lead to transgressions.

1 As you will have appreciated in answering Question 6.1(c), if the rate of production of new oceanic crust increases for any reason, there will be greater amounts of hot and therefore less dense crust standing above the general level of the surrounding deep ocean floor, and this will displace some of the water in the ocean basins onto the surrounding continents. Such an explanation has been cogently argued to account for the world-wide transgression that characterized the Upper Cretaceous (*c*. 90 Ma ago) when sea-level was some 300–400 m higher than today (see Figure 6.17). At that time, shelf seas covered a considerably greater area than today.

2 During periods of continental break-up (Figure 3.1), the overall elevation of the continents is reduced by localized crustal thinning (Figure 3.2) and increased erosion. This erosion adds sediment to the ocean basins, and so a certain amount of water is displaced. The result is a rise in sea-level. Thus, a prolonged period of world-wide continental erosion of the sort likely to be associated with continental dispersal will probably lead to a marine transgression.

Conversely, the collision of continents (as occurred during the formation of the Alps and Himalayas) causes great thicknesses of sediments along former continental margins to become stacked up, and continental blocks to be thickened and isostatically elevated. The result is a fall in sea-level (regression) as ocean waters withdraw into the slightly larger and deeper basins that remain.

Table 6.2 Major Ice Ages in the last part of the geological record.

Period	Approx. time of peak
Quaternary	within the past 2 Ma
Permo-Carboniferous	*c.* 260 Ma
Late Ordovician	*c.* 440 Ma
Late Precambrian	*c.* 650 Ma
Upper Proterozoic	*c.* 900 Ma

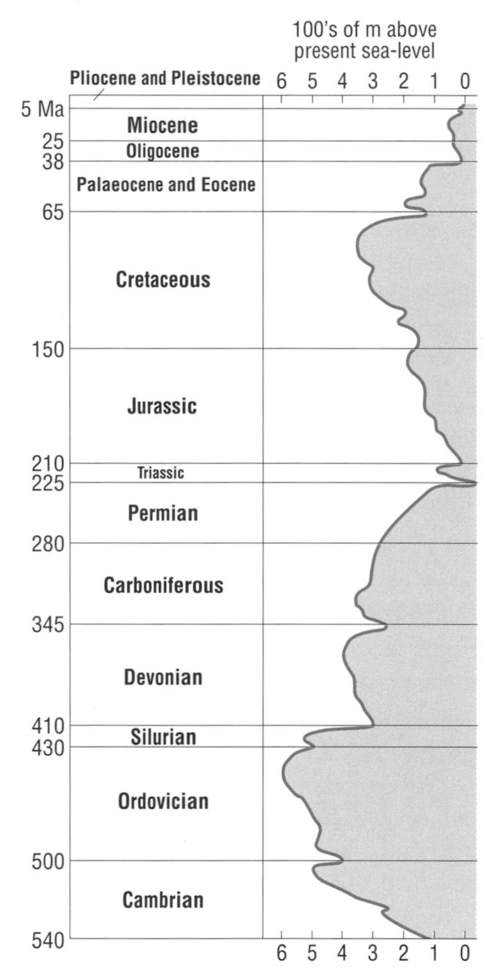

Figure 6.17 A graph of estimated world-wide sea-level since the start of the Cambrian Period.

6.2.9 MAJOR TRANSGRESSIONS AND REGRESSIONS

Marine transgressions and regressions (advances and retreats of the sea over the land) that occurred before the Quaternary cannot be docmented in the same detail as those that occurred during it, because the sedimentary record is too fragmentary. However, there have been five well-documented Ice Ages over the past 900 Ma (Table 6.2), and there is no reason to suppose that the other four were not accompanied by the same kind of short-term glacial–interglacial fluctuations that characterized the Quaternary Ice Age. In addition, there have been larger longer-term marine transgressions and regressions that have taken place over millions rather than tens or hundreds of thousands of years, and which evidently have nothing at all to do with Ice Ages. Classic examples are the late Cambrian (*c.* 520 Ma ago) and Cretaceous (90 Ma ago) transgressions.

One interpretation of world-wide (eustatic) sea-level over the past 540 Ma, based primarily on the sedimentary record, is shown in Figure 6.17. There are many local (isostatic) effects which may influence the picture, so you should not pay much attention to apparent short-term fluctuations, but concentrate on the main trends, which can be assumed to apply globally.

QUESTION 6.12 Examine Figure 6.17 and Table 6.2. Is there any obvious evidence that the major glaciations listed in Table 6.2 could have exerted a dominant control on the trends in sea-level revealed by Figure 6.17?

It is noteworthy that the high sea-level in the Cretaceous coincided with the time when the break-up of Pangea was in full swing (see Figure 3.1) and we might expect a larger proportion of the ocean basins to have been occupied by young, hot, ridges (Figure 6.18). Furthermore, the eruption of lavas forming the large Ontong–Java Plateau occurred during the Cretaceous (Section 4.2.5). When there is an increase in spreading rates, or new spreading axes form, or large igneous provinces develop, then sea-level rises (Question 6.1(c)). The high sea-level during the Ordovician (Figure 6.17) did not correspond to the break-up of a supercontinent, and may be related to an increase in spreading rate or to the formation of now-vanished large igneous provinces.

In the last Chapter of this Volume, we shall briefly examine other long-term aspects of the interaction between seawater and oceanic crust.

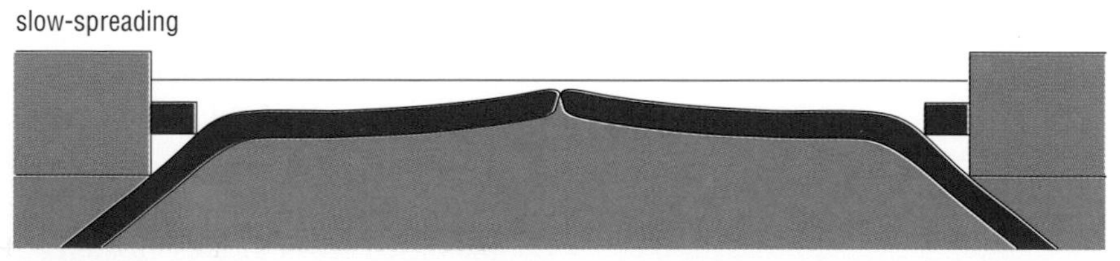

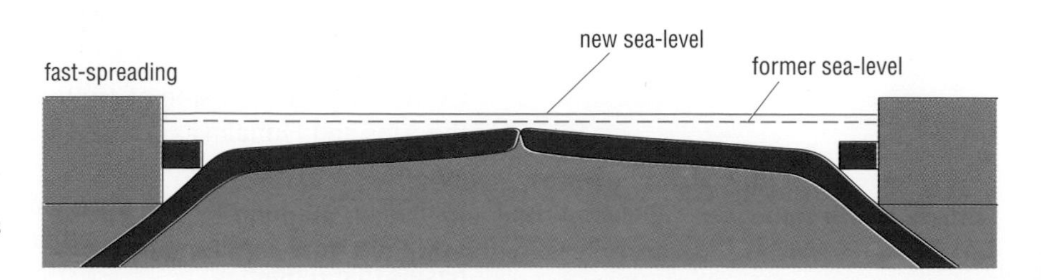

Figure 6.18 Cross-sections (exaggerated, and not to scale) showing how the greater volume of young buoyant oceanic crust during times when sea-floor spreading is fast can lead to higher global sea-level than during periods of slow-spreading. (A similar rise in sea-level can occur independently of spreading *rate* if the total length of spreading axis increases, because this will also increase the volume occupied by spreading ridges and thereby displace seawater.)

6.3 SUMMARY OF CHAPTER 6

1 Sediments of oceanic seismic layer 1 thicken from a few metres maximum at ridge crests to several kilometres at continental margins. Deep-sea pelagic sediments consist of clays and biogenic sediments formed of calcareous and/or siliceous skeletal remains of mostly planktonic organisms. Calcareous sediments tend to predominate on the flanks of ocean ridges, and siliceous sediments and clays accumulate in deeper parts of the ocean basins. Metalliferous sediments resulting from hydrothermal activity are believed to form the base of the sediment sequence nearly everywhere.

2 Sediments and the organisms preserved in them provide much information about the nature and timing of events in the evolution of ocean basins. For example, the spread of a species of the planktonic foraminiferan genus *Guembelitria* in the Southern Ocean has helped to refine our knowledge of the history of continental separation in this region and the development of the Antarctic Circumpolar Current.

3 World-wide (eustatic) changes in sea-level are caused by changes in the total volume of water in the oceans, and by changes in the shape and volume of the ocean basins. Sea-level is presently rising world-wide, as a result of climate warming after the last glaciation, and is beginning to be exacerbated by warming caused by an enhanced greenhouse effect because of anthropogenic CO_2 emissions into the atmosphere.

4 Isostatic effects such as sediment loading or rebound after disappearance of ice-sheets complicate the study of sea-level changes on a local scale. However, satellites can now provide methods for global monitoring of present-day changes in sea-level and sea-surface temperature, and the value of these methods will increase as longer time-series of data are accumulated.

5 Sea-level changes prior to the last glaciation can be reliably determined by measuring $\delta^{18}O$ values in the carbonate skeletal remains of marine organisms. These values are higher when the ice-caps are large, because water vapour (and hence the snow which builds the ice-caps) is relatively enriched in $H_2{}^{16}O$. When ice-caps are small the oceans are comparatively enriched in $H_2{}^{16}O$, and so $\delta^{18}O$ values in marine organisms are lower. It is possible to calibrate $\delta^{18}O$ values in terms of sea-level changes, and water temperatures. This has enabled the growth of the Antarctic ice-sheet to be charted. Growth of the ice-sheet became especially rapid during the Miocene.

6 In the late Miocene (Messinian), a combination of tectonic forces and falling sea-levels resulting from the growth of the Antarctic ice-sheet led to the isolation of the Mediterranean from the world ocean. Seawater in the Mediterranean evaporated and deposits of evaporite salts were laid down. During the main period of evaporite deposition from about 5.5 to 4.8 Ma ago there were many alternations of flooding and desiccation, so that the total thickness of evaporites is greater than 1 km. The barrier at Gibraltar was finally submerged at about 4.8 Ma ago, and present-day conditions were established.

7 The growth and decay of ice-caps is accompanied by expansions and contractions of climatic belts, especially expansion of the cold polar zones. The extent of these fluctuations can be monitored by the study of organisms in deep-sea sediments.

8 Plate tectonic processes are major long-term influences on sea-level. Rapid sea-floor spreading and continental fragmentation and/or formation of large submarine igneous provinces lead to reduction in the total volume of the ocean basins, resulting in widespread transgressions onto continents. Continental collisions result in thickening and uplift of parts of the continental crust, causing sea-level to fall. Deposition of thick accumulations of sediment along continental margins also contributes to sea-level rise.

Now try the following question to consolidate your understanding of this Chapter.

QUESTION 6.13 Which of the following statements are true, and which are false?

(a) The total range of fluctuation in sea-level as a result of growth and decay of ice-caps and glaciers is of the order of 150–200 m.

(b) Radar altimeter data from the *TOPEX/Poseidon* satellite prove that there is a long-term rising trend in sea-level.

(c) Remains of *Guembelitria* in marine sediments enabled the full development of the Antarctic Circumpolar Current to be precisely dated as occurring at 20 Ma ago.

(d) Evaporite salt sequences are always formed at stages 2 and 5 of the ocean-basin life cycle (Table 3.1).

CHAPTER 7 THE BROADER PICTURE

In this final Chapter, we attempt to stand back and take a global perspective. The evolution of the ocean basins cannot be separated from that of the ocean waters which fill them, which in turn depends on conditions in the atmosphere above and on the continents bordering them. Changes in the shape, size and distribution of the ocean basins combine with climatic changes to influence the circulation and even the composition of seawater. These in turn affect the nature and distribution of biological activity and the types of sediments deposited on the ocean floor. Thus, Figure 6.1 applies only to the present-day oceans, and the pattern is subtly but inexorably changing with time.

7.1 THE GLOBAL CYCLE

Chemical elements have been cycled through land, sea and air ever since the Earth acquired its fluid envelope (Figure 7.1), which was not far short of four billion years ago. The mantle is the fundamental source of crustal rocks, but Figure 7.1 shows that it also contributes directly to the oceans and atmosphere by outgassing, i.e. through the expulsion of juvenile volatile constituents (water, carbon dioxide (CO_2), hydrogen chloride (HCl), sulphur dioxide (SO_2) and many other gases), which are products of volcanic activity. These gases can be recycled back through the mantle if they are trapped or chemically combined in rocks or sediments of the oceanic crust and carried down in subduction zones.

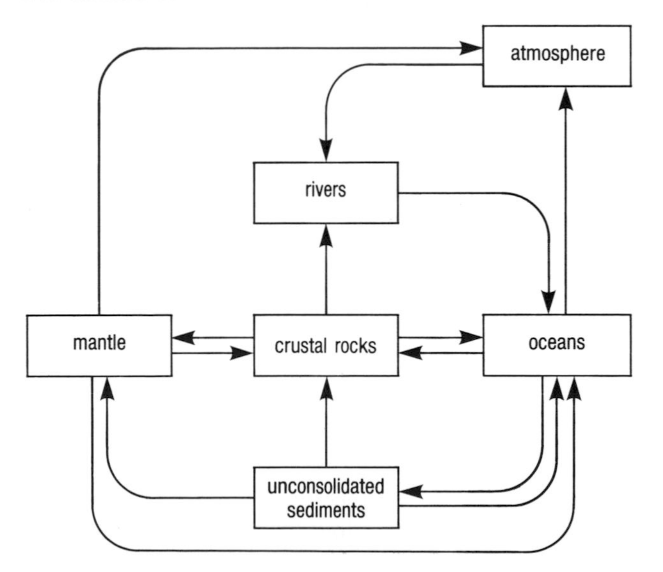

Figure 7.1 The global cycle, involving crust and mantle as well as oceans and atmosphere, illustrating the pathways eventually taken by the elements that pass through the oceans.

The cycle from continent to seawater to ocean floor and sediments, and eventually back to continent can be summarized in four stages, as described below (see also Figure 7.2). At each stage, there may also be direct or indirect chemical exchange with the atmosphere.

Stage 1: Weathering of rocks involves the removal of CO_2 from the atmosphere by rain, and releases cations such as H^+, Ca^{2+}, Na^+, and K^+ and Mg^{2+} into solution in rivers, along with HCO_3^- (bicarbonate) as the dominant anion.

The presence of dissolved CO_2 makes rainwater slightly acidic:

$$CO_2 + H_2O \longrightarrow \underbrace{H_2CO_3}_{\text{(carbonic acid)}} \longrightarrow H^+ + HCO_3^- \qquad (7.1)$$

An important weathering reaction in sedimentary rocks is:

$$\underbrace{CaCO_3}_{\text{(calcite, a common mineral in sedimentary rocks)}} + \underbrace{CO_2 + H_2O}_{\text{(in rainwater)}} \longrightarrow \underbrace{Ca^{2+} + 2HCO_3^-}_{\text{(in solution)}} \qquad (7.2)$$

An example of the type of weathering reaction important in igneous and metamorphic rocks is:

$$\underbrace{2NaAlSi_3O_8}_{\text{(albite, a common mineral in igneous and metamorphic rocks)}} + \underbrace{2CO_2 + 3H_2O}_{\text{(in rainwater)}} \longrightarrow \underbrace{Al_2Si_2O_5(OH)_4}_{\text{(kaolinite, a clay mineral which remains in the weathered rock)}} + \underbrace{4SiO_2}_{\text{(silica, partly in solution)}} + \underbrace{2HCO_3^- + 2Na^+}_{\text{(in solution)}} \qquad (7.3)$$

Other reactions of the general form acid + base = salt + water are common in weathering. The acid is provided mainly by CO_2 in solution. CO_2 is the fourth most abundant gas in the atmosphere (after nitrogen (N_2), oxygen (O_2) and argon (Ar)), and is maintained there by volcanism, the respiratory activity of organisms, and the decay and combustion of organic matter. The bases in the reactions above are supplied by the rocks, and the salts are carried away in solution, in the form of their constituent anions and cations.

What are the principal anions available in seawater to balance the cations there, and what is their source?

They are Cl^- and SO_4^{2-} (Table 5.2), which are products of mantle outgassing by volcanic activity (as HCl and SO_2). These gases are dissolved in rainwater and thereby removed from the atmosphere, and enter the oceans either directly or by way of rivers. In river water, bicarbonate is the dominant anion, as Equations 7.1 to 7.3 indicate. Its removal from solution in seawater is explained in the next stage.

Stage 2: Dissolved constituents are removed from solution in seawater into sediments by a variety of processes. Among the more important are biogenic precipitation of calcium carbonate and silica in the skeletal parts of organisms (removing Ca^{2+}, HCO_3^- and SiO_2 from solution), and the mutual attraction between the charges on ions (especially cations) and the residual charges on sinking particles (such as clay particles). The latter process is known as **adsorption**, and is an important mechanism for removal of dissolved ions (especially of metals such as cobalt, manganese and iron) from solution.

Stage 3: We have seen in Chapter 5 how sea-floor weathering and hydrothermal activity are highly effective mechanisms for removing various cations from seawater, and adding other cations and gases in their place. Other reactions that change dissolved constituents in seawater involve the sediments.

Diagenesis may be defined as the reactions that occur during the consolidation of sediment to form rock (e.g. when siliceous and calcareous deposits become chert and limestone, respectively). In the diagenesis of clays, Mg^{2+} and K^+ are removed from solution in pore waters and incorporated into the clay minerals. The concentration of these elements is maintained in the pore waters by downward diffusion from overlying seawater.

Like hydrothermal activity, diagenesis supplies certain dissolved constituents as it removes others (Section 5.2.2). While some elements are precipitated from sediment pore waters into sediments, others are dissolved from the sediments into the pore waters, and eventually diffuse out into the main body of the oceans.

The cycle can now be completed:

Stage 4a: Consolidated sediments and underlying rocks may be uplifted directly back into the weathering part of the cycle (e.g. in ophiolites or uplifted continental margins).

Stage 4b: More commonly, sediments and oceanic crust may *either* be deeply buried and metamorphosed, especially where thick accumulations of continental margin sediments (accretionary prisms, Section 2.2.2) are caught up in continental collisions, *or* they may be subducted and partially melted to form magmas. In both cases, further chemical changes occur, followed by eventual tectonic uplift or volcanic rise back into the weathering part of the cycle (Stage 1).

Figure 7.2 summarizes the four stages described above, except for those Stage 3 processes that supply elements to seawater.

Figure 7.2 Details of the global cycle.

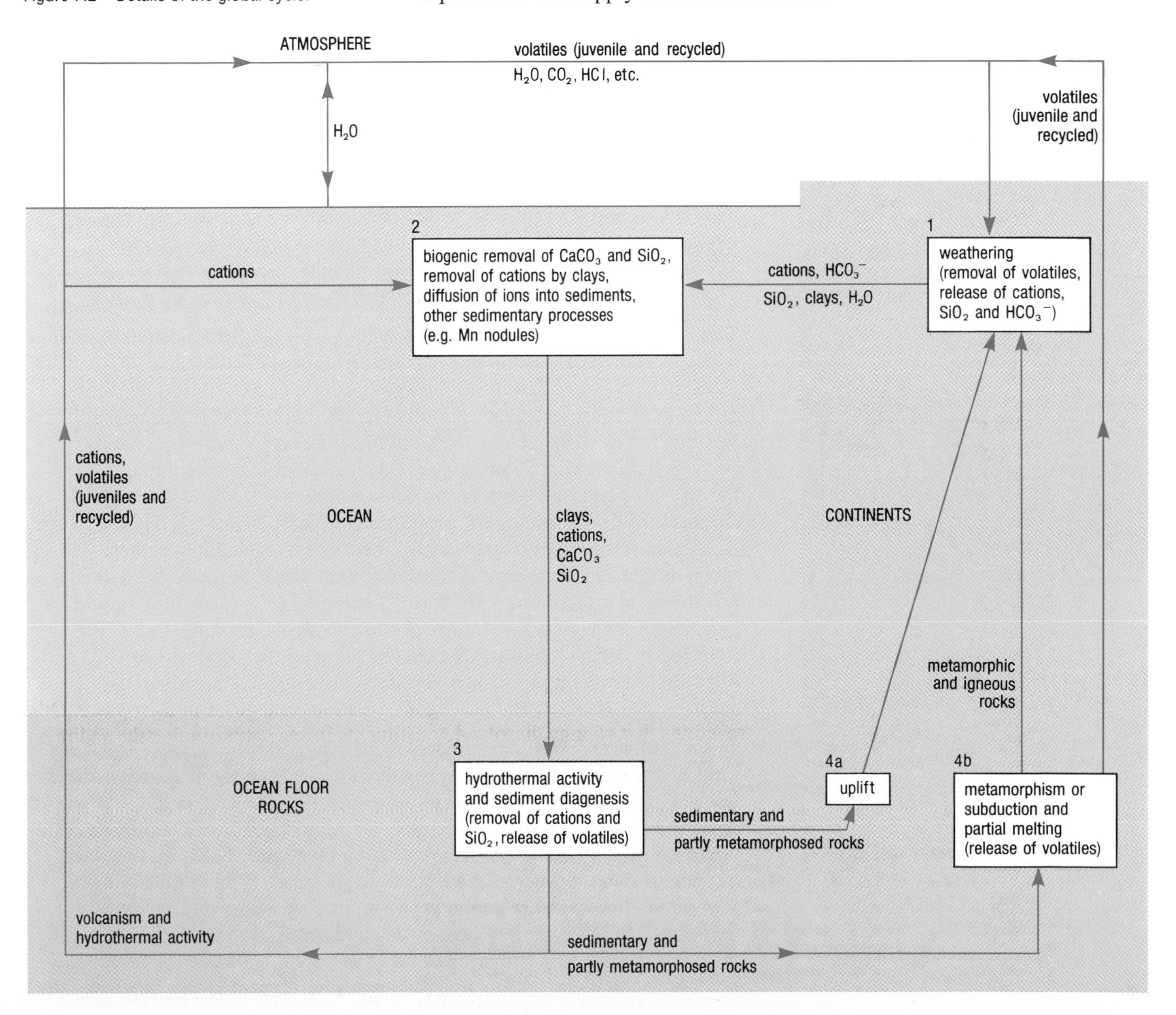

QUESTION 7.1 Would you say that oceanic crust is the same after travelling a few thousand kilometres across the sea-floor 'conveyor belt' as when it leaves its source at the spreading axis? Explain your answer.

7.1.1 CHANGES IN COMPONENTS OF THE CYCLE

Chemical elements have been moving through the cycles of Figures 7.1 and 7.2 (and other related sub-cycles) for about the past 4 billion years. There appear to have been no fundamental changes in the composition of the mantle or igneous crustal rocks during the whole of that time – for instance, basaltic and granitic rocks throughout the Precambrian are virtually identical to basalts and granites of Tertiary age; and the same can be said of most sedimentary rocks. On the other hand, evidence from the sedimentary record suggests that both *oceans* and *atmosphere* were different from what they are now, for much of geological time.

Between about 4.5 and 2.5 billion years ago, the concentration of CO_2 in the atmosphere is estimated to have been several hundred times its present value, and there was virtually no free oxygen. There would thus have been a greater concentration of dissolved CO_2 (and of the ions derived by reactions such as Equations 7.2 and 7.3) in the oceans. The Earth's mantle was hotter in those days, so the rate of generation of new ocean floor by sea-floor spreading was perhaps as much as ten times the rate found in the present-day Pacific Ocean. This would have led to significantly higher rates of hydrothermal activity, and hence of exchange of elements between seawater and rock in Stage 3 of the cycle described above. The absence of free oxygen meant that the early oceans and atmosphere were reducing environments. However, the major constituents of present-day seawater (Table 5.2(a)) were already present, though not in the same proportions.

Thus, 3.5-billion-year-old sedimentary rocks, originally deposited in shallow water, contain calcium and magnesium carbonates ($CaCO_3$ and $CaMg(CO_3)_2$) and calcium sulphate (as gypsum $CaSO_4.2H_2O$), evidently precipitated from seawater. Sodium chloride (halite, NaCl) was probably precipitated in ancient evaporites, but the evidence is more equivocal because halite is very soluble. The reducing conditions meant that most of the sulphur was in the form of insoluble sulphide (S^{2-}) instead of soluble sulphate (SO_4^{2-}) ions, and the proportion of bicarbonate ions (HCO_3^-) must have been greater in the early oceans (because, as noted above, there was so much more CO_2 in the atmosphere). Another consequence of the reducing conditions was that iron, in the form of soluble Fe^{2+}, was a more important constituent than today (when iron occurs mostly as insoluble Fe^{3+}). The Fe^{2+} in the early oceans presumably behaved in much the same way as Ca^{2+} and Mg^{2+} do today, being precipitated in carbonate and silicate minerals.

The atmosphere and oceans probably remained reducing environments until about 2 billion years ago (estimates vary), when the atmospheric $CO_2:O_2$ ratio began to fall markedly. Carbon dioxide concentrations decreased and oxygen began to build up in the atmosphere. The decrease in the atmospheric $CO_2:O_2$ ratio was largely a consequence of the evolution of the biosphere, within which photosynthesizing plants fixed CO_2 and released oxygen. However, oxygen concentrations increased only gradually over geological time, because the oxygen was used up again, both in the oxidation of organic matter (respiration and decay) and in oxidation reactions during the weathering of rocks (e.g. oxidation of Fe^{2+} to Fe^{3+}). It is not known when the present level of atmospheric oxygen was reached, but oxygen has probably been maintained near its present level for about the past 200 Ma.

Evidence that seawater also may have approximated its present composition over the same period comes from the preservation in marine sediments of particular organisms (such as certain echinoderms), whose basic physiology appears incapable of tolerating substantial departures from the range of composition (and temperature) that characterize the modern oceans.

This does not rule out either a gradual change of composition with time or fluctuations about some long-term average composition. In fact, concentrations of some of the major dissolved constituents may have changed by as much as a factor of two or three during the past 500 million years. Evidence of such changes can be found, for example, in evaporites deposited at different times. Thus, evaporites deposited during the Permian, about 250 Ma ago (see Appendix), contain an abnormally high proportion of magnesium salts. Moreover, both the magnesium and the potassium salts are mainly sulphates in these Permian evaporites, whereas chlorides have predominated for most of the past 500 Ma. This has been taken to indicate that Permian seawater had higher concentrations of dissolved magnesium (Mg^{2+}) and sulphate (SO_4^{2-}) than seawater at most other times.

QUESTION 7.2 The Permian was the period when the global supercontinent of Pangea (that you saw breaking up in Figure 3.1) was newly formed. There were no ocean basins separating the continents, so it is likely that the total length of sea-floor spreading axis (which was all within the single global ocean, Panthalassa) was considerably shorter than the total length of spreading axis today. Bearing in mind what you know about the behaviour of Mg^{2+} and SO_4^{2-} in hydrothermal systems, can you suggest why there might have been high concentrations of these ions in Permian seawater (and hence evaporites)?

Apart from the changes in the composition of seawater that have accompanied the gradual evolution of the atmosphere and the cooling of the mantle, and the somewhat faster changes associated with sea-floor spreading patterns, there must clearly be variations in the shorter term. You already know that world-wide climatic fluctuations lead to variations in the amount of water locked up in ice-caps. When the total volume of water in the ocean changes, it must affect the salinity – although the relative proportions of the various dissolved constituents should remain the same. On the other hand, the evaporation of large volumes of seawater from restricted basins leads to the differential precipitation of dissolved constituents.

7.1.2 SOME EFFECTS OF SHORT-TERM CHANGES

We can look at those last two examples in a little more detail, and estimate their effect on the world ocean as a whole. Both can be classified as short-term events, for they have durations of the order of 10^5 years, which are almost trivial when compared with the geological time-scale.

The Messinian event (Section 6.2.6) was a crisis for the Mediterranean; but what was its effect on the world ocean? If we assume for simplicity a total of 1 km thickness of evaporites over about four-fifths of the area of the Mediterranean floor (i.e. $2 \times 10^6\,km^2$) (cf. Question 6.9), that would give us a total volume of $2 \times 10^{12}\,m^2 \times 1 \times 10^3\,m = 2 \times 10^{15}\,m^3$. At an average density of $2 \times 10^3\,kg\,m^{-3}$, that is about 4×10^{18} kg of salts in the Mediterranean evaporites.

QUESTION 7.3

(a) The oceans have 1.4×10^{21} kg of water, and each kilogram contains 35 g of dissolved salts. What is the total mass of salts in the oceans?

(b) What percentage of your answer to (a) is represented by the Mediterranean evaporites?

Clearly, on a global scale, the Messinian salinity crisis caused the removal of a significant fraction of the total salt in the world ocean. However, remember that this removal did not happen all at once: it was spread over a total of about 700 000 years, and the dissolved constituents of salts continued to be carried into the open oceans by rivers.

In Section 6.2.4, you read that the Quaternary fluctuations of sea-level were the result of around $50 \times 10^6\,km^3$ of water being alternately withdrawn from and returned to the oceans.

QUESTION 7.4

(a) Approximately what proportion is that of the total amount of water (1.4×10^{21} kg) in the oceans? (The average density of seawater is $1025\,kg\,m^{-3}$.)

(b) Would you expect the average salinity of seawater to be greater or less than at present during a glacial period?

7.1.3 THE STEADY-STATE OCEAN

Evidence from different parts of the global cycle (Figures 7.1 and 7.2) has led marine scientists to conclude that for most dissolved constituents the oceans can be regarded as being in a **steady-state** condition when considered over long timescales. This is because the rates of input and removal of the majority of dissolved constituents are generally in balance over periods that are very long (millions of years) compared with those necessary for oceanic mixing (about a thousand years).

In other words, on a timescale of a few to several million years, the rates of removal processes such as biological precipitation, diagenesis and metamorphism (Stages 2 and 3 of Figure 7.2) are in balance with those of supply processes such as weathering, volcanism and hydrothermal activity (Stage 1 and parts of Stages 3 and 4). The implication is that, on average, for every mole of, say, calcium, magnesium, potassium or sulphate removed from solution in seawater, another mole is added from rivers or hydrothermal solutions. However, when there is a change in the rate of supply or removal of a dissolved constituent, it can take time for a steady state to be re-established. Thus, in Question 7.2 we inferred that an increased concentration of dissolved magnesium in Permian seawater was caused by a reduction in the rate of removal of magnesium from solution in seawater. At the same time, the removal rate of magnesium from seawater into evaporites was increased, which tended to compensate for this. A steady state was probably eventually re-established during the Permian, when the summed rates of magnesium removal from seawater into evaporites and into rocks by hydrothermal exchange came into balance with the rate of supply from rivers (which for simplicity we assume not to have changed, although we have no proof). However, when this had been achieved, the concentration of magnesium in seawater must have been higher than its previous value. After the Permian, when Pangea broke up and hydrothermal exchange had re-established itself at a faster rate, the magnesium concentration of seawater declined, the rate of formation of magnesium-enriched evaporites fell with it, and a new steady state was established.

It should also be clear from Figures 7.1 and 7.2 that we cannot confine ourselves simply to the balance between processes of supply and removal of dissolved constituents. Figure 7.2 shows that volatile constituents escaping from the mantle continue to be added to the cycle, notably H_2O, CO_2 and HCl in volcanic gases. Igneous rocks originating from the mantle and erupted at spreading axes affect the composition of seawater through sea-floor weathering and hydrothermal activity. It is difficult to know just how much of this contribution from the mantle is really 'new' and how much is being recycled (over periods of up to hundreds of millions of years) from sediments and crustal rocks (plus seawater trapped within them). Such recycling would occur following subduction into the mantle at destructive margins. The actual *net* addition of new constituents, however, is generally considered to be very small in comparison with the amounts being circulated in the cycles shown in Figure 7.2.

The long-term geochemical stability of the Earth's outermost layers is particularly impressive when looked at it in the context of the obvious physical and biological changes that have affected the shape and appearance of continents and ocean basins.

QUESTION 7.5 How might the steady-state ocean respond (a) to a decrease in the supply of dissolved Ca^{2+} from rivers, and (b) to an increase in the supply of dissolved Mg^{2+} from rivers?

7.2 SOME RATES COMPARED

In this final Section, it is useful to look back and make some comparisons between the rates of the various processes you have been studying.

Rates of formation of oceanic crust and plate movement are very slow compared with the rates at which hydrothermal systems operate. At a modest spreading rate of 2 cm yr^{-1}, a piece of newly formed oceanic crust will have moved 200 km in 10 Ma. In that time, the equivalent of the entire volume of the ocean waters will have circulated through the oceanic ridge system, through literally thousands of black smokers and white smokers and low temperature systems, the lifespan of each of which can be no more than a few tens of thousands of years at most (Section 5.3.3).

Rates of reaction in hydrothermal systems are very rapid: in laboratory experiments where seawater and basalt are heated together, the complete removal of magnesium from seawater into basalt is achieved in a matter of a week or two at 300 °C, although it takes several months at lower temperatures.

These high rates of reaction ensure that any significant change in the conditions under which the reactions occur should be rapidly reflected in the hydrothermal solutions emanating from spreading axes (Figure 5.8). However, such changes are rare, and as we saw in Section 5.3.2 the composition and temperature of most vent solutions have been observed to be stable over periods exceeding ten years. Longer-term stability is to be expected as well.

The vent solutions emerge with velocities that range from tens of centimetres to metres per second. However, those rates of motion apply only to the narrow upflow zone of hydrothermal plumbing systems (Figure 5.5).

Water probably moves through the much broader downflow zone at rates comparable with the advance of the cracking front, i.e. not more than a few centimetres per day. This is consistent with the periods of time required for the chemical reactions to reach equilibrium in hydrothermal systems.

When we compare the rates at which these crustal processes operate with rates of global sea-level fluctuation, we find that rates of sea-level change are not greatly different from those of plate movements. For the past few thousand years, the average rate of rise in global sea-level has been 1–2 mm yr^{-1}, but a glance at Figure 6.10 shows that this is atypical, at least so far as glacial–interglacial intervals are concerned.

Can you see that over much of the time since the last glacial maximum sea-level has risen at a rate of almost 1 cm yr^{-1}?

Figure 6.9 shows that sea-level rose about 80 m between about 17 000 and 7000 years ago, which is about 8×10^{-3} m yr^{-1}, or 8 mm yr^{-1}.

Localized or isostatic changes of sea-level can be much more rapid: the formation of a raised beach, for example, can involve vertical movements of several tens of cm per year – these result from tectonic forces along particular parts of continental margins.

Most of this Volume on ocean basins has been concerned with ‘solid Earth’ components of Figures 7.1 and 7.2, but it should already be obvious that the oceans are a very important part of the global cycle. Other Volumes in this Series are concerned principally with sub-cycles that involve the oceans and the atmosphere; although interactions with the solid Earth are, of course, considered when appropriate.

7.3 SUMMARY OF CHAPTER 7

1 The oceans are an important part of the global cycle of chemical elements. Cations (positively charged ions) are supplied to the oceans mainly as a result of atmospheric weathering of terrestrial rocks, and partly from hydrothermal activity. Anions (negatively charged ions) to balance these in the freshwater environment come from the atmosphere (mainly CO_2 as bicarbonate HCO_3^- ions in solution), but in the oceans the cations are balanced mainly by chloride and sulphate anions, whose principal source is volcanic outgassing from the Earth’s interior.

2 A variety of biological and chemical processes are responsible for removal of dissolved constituents from seawater. The resulting sediments and metamorphosed crustal rocks may eventually be returned to the continents by some combination of accretion, subduction, collision, metamorphism, volcanism and uplift.

3 The compositions of the atmosphere and oceans 2.5–4.5 billion years ago were markedly different from what they are today. Even as recently as about a billion years ago, the environment was more reducing, because there was less oxygen and more CO_2. Evolution of the biosphere has played a major role in decreasing the $CO_2 : O_2$ ratio, and the atmosphere has had approximately its present composition for about the past 200 million years. There may have been variations in seawater composition associated with changes in the rate of hydrothermal exchange, or the rate at which weathering products have been supplied to the oceans from continents.

4 The oceans are believed to be in a steady-state condition when measured over periods of millions of years, the supply and removal of dissolved constituents being in overall balance.

5 The speed of movement of water through hydrothermal systems is orders of magnitude greater than the rate of movement of lithospheric plates. Huge volumes of water are cycled through oceanic crust and huge quantities of chemical elements are exchanged in the time that it takes for a piece of crust to move a couple of hundred km from the ridge axis. Rates of global sea-level rise and fall may at times be as great as those of horizontal plate movements.

Now try the following questions to consolidate your understanding of this Chapter.

QUESTION 7.6 Direct measurements of the permeability of the rubbly layer 2A of oceanic crust indicate rates of 'groundwater' movement of approximately $6 \times 10^{-7}\,\mathrm{m\,s^{-1}}$. How long does it take water to move a distance of 1 m through the rocks?

QUESTION 7.7 In Section 5.5 you read that over 10% of the calcium supplied to the oceans comes from hydrothermal sources. Yet the amount of water circulating through ocean ridges (*c.* 10^{14} kg) each year is less than 0.5% of the annual supply of river water to the oceans (*c.* $300 \times 10^{14}\,\mathrm{kg\,yr^{-1}}$) that provides most of the rest of the calcium. Can you explain the apparent paradox, i.e. how can less water supply more calcium?

APPENDIX THE STRATIGRAPHIC COLUMN

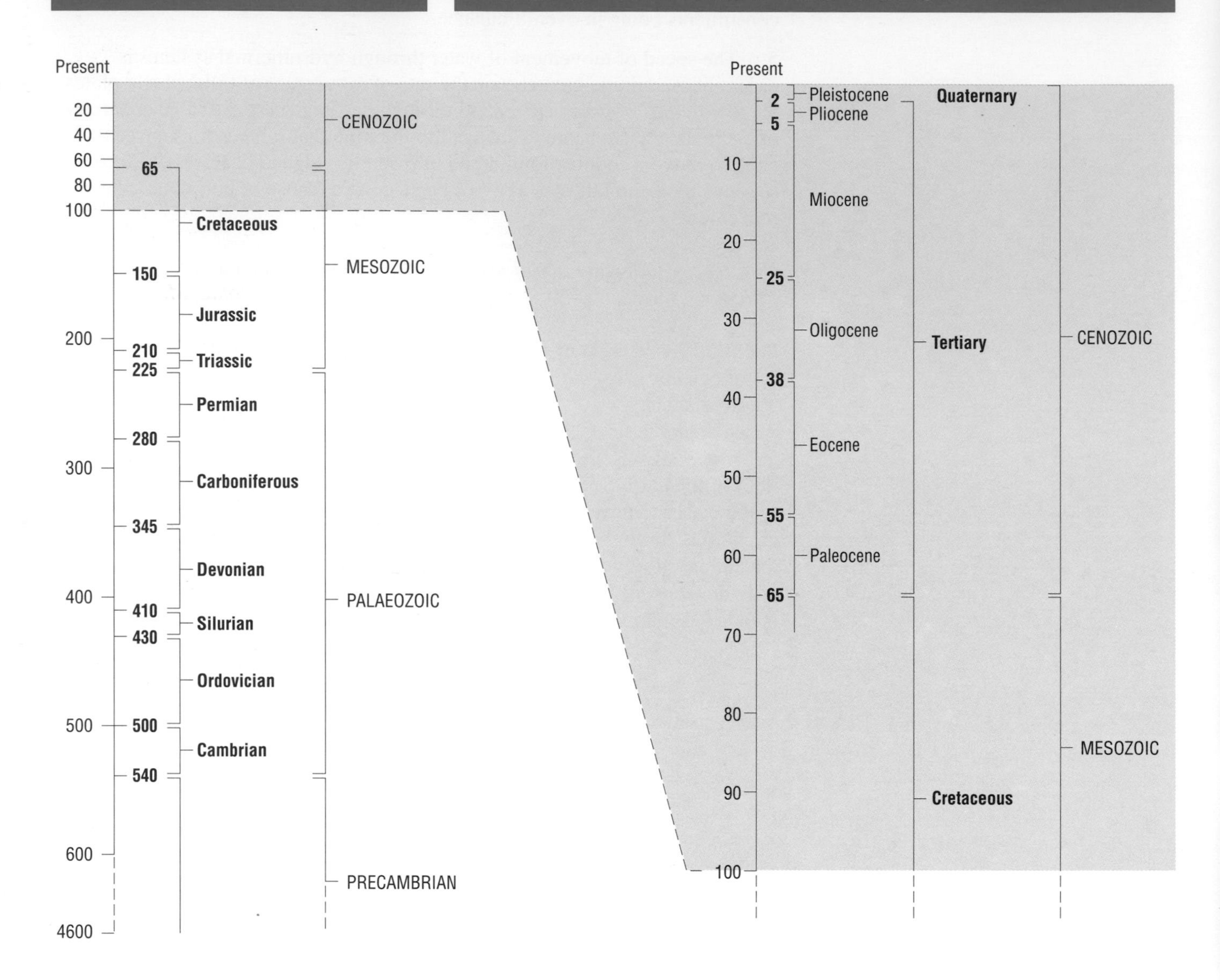

Appendix The stratigraphic column. Ages are given in millions of years (Ma).

SUGGESTED FURTHER READING

HUMPHRIS, S. E., ZIERENBERG, R. A., MULLINEAUX, L. S. AND THOMSON, R. E. (eds) (1995) *Seafloor Hydrothermal Systems*, Geophysical Monograph 91, American Geophysical Union. A collection of review papers about hydrothermal circulation, vents, plumes, and the biota that depend upon them.

KAHARL, V. A. (1990) *Water Baby – the story of Alvin*, Oxford University Press. A journalistic (but scientifically adept) account of the first 25 years of deep-sea operations using the submersible *Alvin*.

MacLEOD, C. J., TYLER, P. AND WALKER, C. L. (1996) *Tectonic, Magmatic, Hydrothermal and Biological Segmentation at Mid Ocean Ridges*, Special Publication No. 118, Geological Society. A collection of chapters summarizing new research.

MENARD, H. W. (1986) *The Ocean of Truth: A Personal History of Global Tectonics*, Princeton University Press. The story of marine geological exploration during the 1950s and 1960s which led to the plate tectonics revolution, and the scientists who participated.

PARSON, L. M., MURTON, B. J. AND BROWNING, P. (eds) (1992) *Ophiolites and their Modern Oceanic Analogues*, Special Publication No. 60, Geological Society. A collection of chapters summarizing new research, intended for a geological readership.

PIRIE, R. G. (ed.) (3rd edn, 1996) *Oceanography: Contemporary Readings in Ocean Sciences*, Oxford University Press. A collection of 37 specialist chapters covering much of the scope of this Series (and beyond). Chapters relevant to this Volume deal with exploration, sea-level rise, global warming, seamounts and the ecology of hydrothermal vents.

SUMMERHAYES, C. P. AND THORPE, S. A. (eds) (1996) *Oceanography: An Illustrated Guide*, Manson Publishing. An introductory book, covering much of the scope of this Series as a whole, at advanced undergraduate level.

ANSWERS AND COMMENTS TO QUESTIONS

CHAPTER 1

Question 1.1 The reliability with which *positions* can be determined depends mainly upon the navigational technique used. The eastern continental shelf of the United States is about 150 km wide, so the edge of the continental shelf is beyond the range of line-of-sight navigational methods. If either GPS or a radio navigation system were used (Table 1.1), the positional errors for both (a) and (b) would be in the order of 100 m, which is not important on the horizontal scale of Figure 1.8. Positional errors would be reduced to only about 10 m if differential GPS had been available. However, if only celestial navigation were available, then the position of both features could be out by as much as a few km.

To obtain *depth* data, time differences have to be converted into distances, and the principal potential for error is in using the wrong sound velocity for the conversion. Moreover, low-resolution echo-sounders will not always give reliable results in areas of variable topography (Figure 1.9(b)). However, (a), the edge of the continental shelf is relatively featureless and the water is shallow, so that the profile is probably reliable to within a few metres.

When we consider (b), the 'bump' on the continental slope, we have to appreciate the great vertical exaggeration of Figure 1.8. This feature is actually about 1 km across and not much more than 50 m high, so the slopes are fairly gentle. It seems probable that here also the error is unlikely to be more than a few metres.

Question 1.2 Changes that result from the behaviour of varying winds, currents, tides and so on will eventually average out if the satellite repeats its track every few days (preferably at intervals that are not multiples of the tidal cycle).

Question 1.3 The horizontal scale suggests that the wavelength of geoid variations, as deduced from sea-surface height, is of the order of several tens to a few hundreds of km. The vertical scale suggests that the amplitude of these variations is between about 1 and 2 m.

CHAPTER 2

Question 2.1 (a) About 70–71% of the Earth's surface lies below sea-level, according to both the histogram and the hypsographic curve.

(b) The scale of the hypsographic curve shown in Figure 2.4(b) makes it impossible to tell precisely, but a rise in sea-level of only 100 m would flood an extra 5–10% of the Earth's surface. There would be a drastic and dramatic shift in the position of shorelines, which are ephemeral features even on an historical time-scale and are much more so in terms of geological time.

(c) The total vertical relief is 8.85 km + 11.04 km, from the hypsographic curve. This is only about 0.3% of the Earth's mean radius (which corresponds roughly to the sort of relief you find on the skin of an orange).

Question 2.2 (a) The principle of isostasy requires that thinned continental crust 'floats' lower than thick crust. An analogy with wood blocks in water is appropriate: the top of a thin wooden block will be closer to the water surface than that of a thick block.

(b) The transition between continental and oceanic crust is clearly not at the coastline. It must lie somewhere beneath the region between the edge of the continental shelf and the upper part of the continental rise.

Question 2.3 (a) (i) The continental shelf is consistently quite narrow, rarely more than 50 km across. The shelf break seems to be gradual or abrupt (as with aseismic margins), but (ii) the continental slope is convex throughout its considerable vertical extent of up to 8 km. The continental rise is missing completely, because of the trench at the foot of the slope.

(b) In general, continental slopes are steeper than along aseismic margins, though the upper part of profile 1 has a gradient of only 1 in 35. The slope in profile 2 is about 1 in 8 (*c*. 7°).

(*Note:* It is important to bear in mind the great vertical exaggeration in these profiles; the vertical exaggeration in Figure 2.7 is × 50 and in Figure 2.9 it is × 25.)

Question 2.4 Successively *younger* slices of sediment are scraped from the descending plate and are added to the *bottom* of the pile making up the inner trench wall. It follows that the oldest slices are at the top of the accretionary prism.

Question 2.5 (a) The Mid-Atlantic Ridge is profile (a), and the East Pacific Rise is profile (b).

(b) The Mid-Atlantic Ridge obviously has much rougher surface relief.

(c) The average gradients are less than 1°, which is generally considerably less than those of continental margins, except for the continental shelf (Figure 2.7 and related text).

(d) The Carlsberg Ridge should resemble profile (a) more than (b), because its spreading rate is similar to that of the Mid-Atlantic Ridge.

Question 2.6 Figure 2.19 shows a clear shelf break, a continental slope with a gradient of about 1 in 100, and a well-defined continental rise, beginning about 200 km from the shelf break at about 2 km depth and grading down into the abyssal plain some 600 km out and 4 km down. The plain itself grades into abyssal hill topography at about 1500 km from the shelf edge and about 5.5 km depth. The aseismic margin of a major ocean is indicated. Therefore, the most obvious answer is the Atlantic – and in fact, this profile is of the Cape Verde abyssal plain and the continental margin off West Africa.

Question 2.7 (a) 3400 km in 43 Ma will give an average rate of movement of:

$$\frac{3400 \times 10^3\,\text{m}}{43 \times 10^6\,\text{yr}} = 0.079\,\text{m yr}^{-1} \approx 7.9\,\text{cm yr}^{-1}$$

(b) At the same rate of movement, the age at the northern end of the chain would be approximately:

$$\frac{1900 \times 10^3\,\text{m}}{7.9 \times 10^{-2}\,\text{m yr}^{-1}} + 43\,\text{Ma} = 24\,\text{Ma} + 43\,\text{Ma} = 67\,\text{Ma}$$

(c) The kink could be explained by a change in the direction of movement of the Pacific Plate at about 43 Ma ago. However, more data from these and other chains would be needed for confirmation.

(d) The duration of the period concerned is 72 Ma – 43 Ma = 29 Ma. To account for 1900 km of movement during this period, the average rate would have to be:

$$\frac{1900 \times 10^3\,\text{m}}{29 \times 10^6\,\text{yr}} = 0.066\,\text{m yr}^{-1} = 6.6\,\text{cm yr}^{-1}$$

Question 2.8 To form a ridge, the hot-spot volcanism would have to be virtually continuous rather than episodic, as in seamount/island chains.

Question 2.9 (a) There are three or four features suggesting segments of a probably aseismic ridge (at about 80° E) parallel with the Ninety-east Ridge. Neither this ridge nor the basins between it and the Ninety-east Ridge fit well with the bathymetric chart in Figure 2.24(b).

(b) The Diamantina Trough is associated with a linear east–west geoid low immediately to the north, which extends far to the east, along the southern margin of Australia, well beyond the eastern end of the trough on the chart.

(c) The correspondence is good, and the features all have geoid anomalies in excess of +2 m.

(d) The more important fracture zones show up as breaks in the otherwise high geoid anomaly along the spreading axes.

(e) The Java Trench appears as a very deep low in the geoid. In fact, though the contours do not make this clear, its maximum depth in the geoid is 20 m. There appears to be a ridge in the geoid, about 3 m high, running parallel to the south side of the trench. This is in fact an artefact, introduced by the filtering technique used to remove long wavelength features in the geoid. It does not represent a real bathymetric ridge. Artefacts are associated with all the real bathymetric features in these geoid anomaly maps, but as their magnitude is only about 15% of that of the associated real feature, they are usually too small to affect the contours at this scale.

Question 2.10 (a) They are the islands (seamounts) of Mauritius and Réunion.

(b) (i) The only fracture zones to be apparent on the *Seasat* data are some crossing the Central Indian Ridge, south-east of the triple junction. There is an offset on the South-west Indian Ridge at about 60° E that suggests a transform fault, but no clear trace of this in the form of a fracture zone beyond the offset. In contrast, the *Geosat*/*ERS*-1 data show fracture zones either side of both ridges, and the offsets to the ridge axis at the sites of transform faults are apparent, notably on the Central Indian Ridge at about 20° S, where the *Seasat* data are particularly ambiguous. (ii) The *Seasat* data show the median valley of the South-west Indian Ridge fairly distinctly, but there is no hint of a median valley on the Central Indian Ridge. In contrast, the *Geosat*/*ERS*-1 data show median valleys quite distinctly on both ridges.

Question 2.11 Slow-spreading ridges are characterized by strong topographic relief compared with fast-spreading ridges (Figure 2.11), reflecting the effect of crustal cooling and subsidence with age (Figure 2.13). For a given distance from a ridge, crust formed at a slow-spreading ridge will be older and therefore deeper than crust formed at a fast-spreading ridge. The geoid in the vicinity of a spreading ridge is correlated with the topography of the ridge: the more pronounced the ridge, the larger the anomaly. In general, therefore, the amplitude of a geoid anomaly over a ridge is inversely correlated with spreading rate. In Figure 2.26, there is a more pronounced geoid anomaly corresponding with the South-west Indian Ridge than with the Central Indian Ridge. The former also has a rougher-looking appearance. On both counts, the former ought to be spreading more slowly than the latter. [*Comment:* This is indeed the case. Mapping of magnetic stripes (cf. Figure 2.13) shows that the half-rate of spreading of this part of the South-west Indian Ridge is 7.5 mm yr^{-1}, whereas that of the Central Indian Ridge is about 25 mm yr^{-1}.]

Question 2.12 The depth of the sea-floor must be 4000 m. This corresponds to an age of about 15–20 Ma, using Figure 2.13. The assumptions that must be made are that the volcano formed at the same time as the portion of oceanic crust upon which it rests, and that local isostatic adjustments can be ignored so that it can be assumed to have subsided at the same rate as the crust.

If the volcano formed at a ridge crest, its summit would have been about 1.5 km below sea-level. This is far below the limits of wave action, so we should not expect a flat top due to erosion. The same argument applies even more strongly if the volcano is younger than the age estimated above.

Question 2.13 (a) False. The age–depth relationship means that since sea-floor of a given age has subsided at the same rate, whatever the spreading rate, then crust generated at a slow-spreading ridge has to travel *less far* before it has subsided to a given depth. Figure 2.11 (and related text) shows that the gradients on fast-spreading ridges are gentler than those on slow-spreading ridges.

(b) Could be either true or false: the evidence is not sufficiently conclusive. We have given only two examples in the text; one supports the model, the other is less clear. It is an attractive idea, but there are insufficient data to be certain even about most linear island chains.

(c) True. As the oceanic crust gets older, more and more sediments accumulate on it.

(d) True. The principle of isostasy (remember the analogy with wood blocks) requires thinned continental crust to be less elevated than continental crust of normal thickness. The only circumstances in which a large area of continental crust of normal thickness could be below sea-level would be if it were forcibly held down in some way.

(e) Partly true. Some sediment is subducted but much is also scraped off onto the inner wall.

Question 2.14 (a) The sea-floor on the north of the fracture zone is deeper and therefore older than that on the south side. Since it is older, it must have spread farther from the ridge at which it was formed. This means that the ridge is offset to the left. Note that the sense of offset is the same whether you imagine yourself looking northwards or southwards along the ridge.

(b) The age–depth relationship shows that the age of sea-floor with a depth of 5000 m is about 47 Ma. We know that this depth is reached at a distance of 940 km from the ridge. As we know both the age and the amount of spreading, we can calculate the average spreading rate:

$$\text{rate} = \frac{\text{amount of spreading}}{\text{age}} = \frac{940 \times 10^3\,\text{m}}{47 \times 10^6\,\text{yr}} = 2.0 \times 10^{-2}\,\text{m yr}^{-1}$$

$$= 2.0\,\text{cm yr}^{-1}$$

(c) Spreading rates of around 2 cm yr^{-1} and less are classified as 'slow'. In terms of its spreading rate, this ridge resembles the Mid-Atlantic, Carlsberg, Central Indian and South-west Indian ridges.

(d) To work this out, we need to know how far the neighbouring sea-floor, north of the fracture zone, is from the segment of the ridge at which it formed. We are told that the depth here is 5500 m. According to the age–depth relationship, this corresponds to an age of about 69 Ma. As spreading rate is almost certain to be identical on both sides of the fracture zone, we can calculate the amount of spreading:

$$\begin{aligned}\text{amount of spreading} &= \text{rate} \times \text{age}\\ &= 2.0 \times 10^{-2}\,\text{m yr}^{-1} \times 69 \times 10^6\,\text{yr}\\ &= 1.38 \times 10^6\,\text{m} = 1380\,\text{km}\end{aligned}$$

So, north of the fracture zone we are at a distance of 1380 km from the ridge, whereas south of the fracture zone we are at a distance of 940 km from the ridge. The offset on the ridge is simply the difference between the two:

$$1380\,\text{km} - 940\,\text{km} = 440\,\text{km}$$

A completed map of the area is shown in Figure A1.

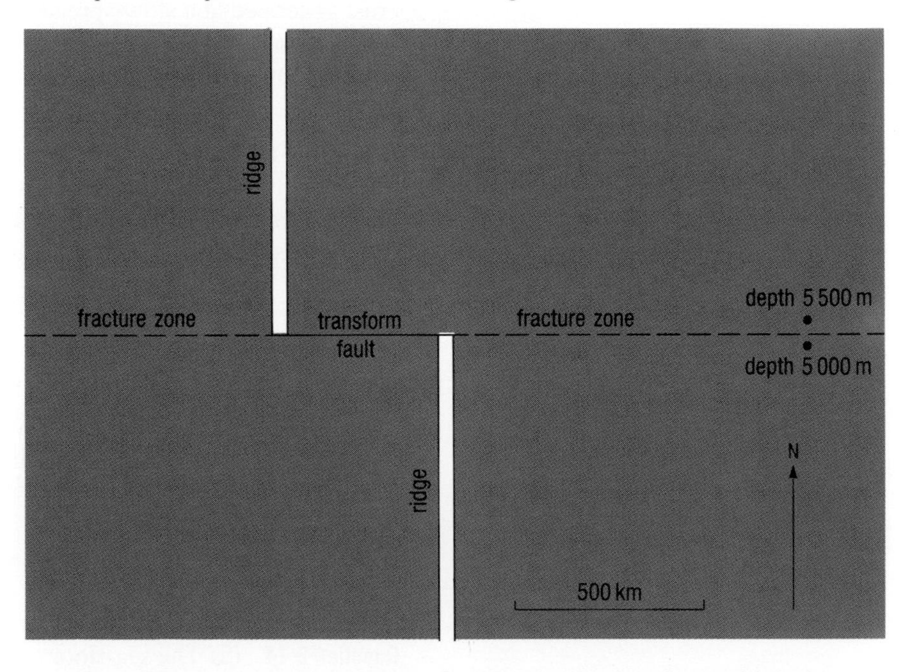

Figure A1 Completed version of Figure 2.29 (for use with answer to Question 2.14).

(e) There is a considerable amount of scatter in the data from which the age–depth curve is derived, and the curve itself is beginning to flatten off as the lithosphere cools, making it difficult to relate depth precisely to age. Based on the age–depth relationship only, there could be errors of ± 5 Ma, or about ± 10%, in the age estimate. The estimate of spreading rate (b) could therefore be in error by at least ± 10%. When we come to *compare* the ages of the sea-floor on either side of the fracture zone, the imprecisions in the age–depth curve become more critical. If the age estimates for the two sites were each out by 5 Ma in opposite senses, then the age difference would be in error by 10 Ma, and the estimate of the ridge offset would be out by 200 km. Magnetic anomaly data would help a lot here.

Question 2.15 The East Pacific Rise is a fast-spreading ridge (Section 2.3.1) and so we would expect it to have relatively smooth topography and a poorly developed or absent median valley (Question 2.5). Figure 2.28 therefore shows exactly what we would expect.

CHAPTER 3

Question 3.1 (a) Tethys has been shrinking since the Jurassic (170 Ma) to the present because of the rotation of Eurasia and Africa towards each other, and the northward movement of India. The Mediterranean Sea and possibly the Black and Caspian Seas are all that is left of Tethys.

(b) It should be apparent from Figure 3.1 that different parts of the Atlantic Ocean began to open at different times. From this Figure, we can see that the southern Atlantic (between South America and Africa) and northern Atlantic (between North America and West Africa) began to open between 170 Ma and 100 Ma ago, but the northernmost Atlantic (either side of Greenland) did not begin to spread until sometime between 100 Ma and 50 Ma ago.

(c) Between 170 and 100 Ma ago, according to Figure 3.1.

(d) 170 Ma to 100 Ma, because fragmentation of Gondwanaland was already well advanced by 100 Ma ago: the Atlantic was open in the south and the Indian Ocean had begun to open as Africa separated from Australia and Antarctica. During the period 100 Ma to 50 Ma ago, India travelled very rapidly northwards towards Asia, Australia began to move away from Antarctica, and Panthalassa, of which the Pacific is the remnant, shrank.

(e) The Atlantic and Indian Oceans continued to expand, and India collided with Asia, causing the formation of the Himalayas. The Pacific continued to shrink, and island arcs developed along its western margins.

Question 3.2 The pole of relative rotation between two plates can be located by reference to spreading ridge segments and transform faults separating the plates (Figure 2.14). In the Red Sea, the axial rift segments are offset from one another but are roughly parallel to the length of the Red Sea (Figure 3.3(a)). The pole must lie somewhere close to the extension of a line running along the length of the Red Sea. We have seen that the opening of the Red Sea appears to be propagating from south-east to north-west, so we can assume that the spreading rate is greatest in the south-east. As spreading rates decrease towards the pole, the pole must lie to the north-west. There are no clearly defined transform faults within the Red Sea, but the fault along the Gulf of Aqaba/Dead Sea line is effectively a transform margin (a conservative plate boundary) and so must lie approximately along an arc of a small circle about the pole. In fact, the pole of rotation is thought to lie somewhere between Sicily and Crete.

Question 3.3 The ocean floor adjacent to North America and west Africa is the oldest in the Atlantic (about 160 Ma old), the ocean floor adjacent to South America and southern Africa is younger (about 110 Ma old), and offshore of

Greenland and north-west Europe the ocean floor is only about 60 Ma old. Thus, as already established in answering Question 3.1(b), the order was (ii), (iii), (i) as follows: (ii) the northern Atlantic began to open first (at this time, North America must have been moving westwards relative to South America, which was still joined to Africa), then (iii) the southern Atlantic began to open, and finally spreading propagated northwards into (i) the northernmost Atlantic.

Question 3.4 (a) The patterns on Figure 3.6 are not easy to unravel, but there is clear evidence of a *southward* increase in age (52 to 65 to 80 Ma) east of the ridge, and a *northward* increase (65 to 80 Ma) west of the ridge.

(b) They were developed well before the present South-east Indian Ridge came into being. Symmetrical spreading about this ridge began about 35 Ma ago, and in a NNE/NE to SSW/SW direction, rather than north–south.

(c) The ridge which generated the crust east of the Ninety-east Ridge must have lain to the north, because ages decrease in that direction. There is no crust younger than 40–50 Ma there, so the ridge must have been subducted down the Java Trench.

Question 3.5 The oldest crust in the Mediterranean is in the east, and it is likely that the thickest sediments are there, too. The western Mediterranean has much younger crust. The older the crust, the deeper it lies (Figure 2.13), so the western Mediterranean should have more 'buoyant' crust than the eastern. If the sea-surface (geoid) mimics the configuration of the ocean floor (Figure 1.17), then we would expect it to be more depressed in the east. This is indeed the case, judging from the geoid anomaly map in Figure 1.18, which shows the eastern Mediterranean in reddish tone (low sea-surface) but the western Mediterranean in blue–green tones (about the global average).

Question 3.6 From Figure 3.6, ocean floor age bands south of India show a greater rate of sea-floor spreading for the period before about 45 Ma ago than afterwards. Therefore the northward rate of movement of India must have been greater before 45 Ma ago than after.

Question 3.7 By simple proportion, using a ruler, the ratio of width of sea-floor from 0 to 52 Ma for the East Pacific Rise to that of similar age for the Mid-Atlantic Ridge is close to 3.3. If the average spreading rate for the Mid-Atlantic Ridge was 2 cm yr^{-1} over this period, then for the East Pacific Rise it was 2×3.3 or nearly 7 cm yr^{-1}. (*Note:* These are spreading rates per ridge *flank*; the total spreading rate will be double these values.)

CHAPTER 4

Question 4.1 The speed of sound in water is *c.* 1.5 km s^{-1} (Section 1.1.2), which is less than that in rock (Figure 4.1). Water will occupy fractures in rocks on the sea-bed, and so will reduce the speed of sound (the seismic velocity) through fractured rocks relative to unfractured rocks, in which there is no water to 'retard' the sound waves.

Question 4.2 In the case of the Mid-Atlantic Ridge, if the half-spreading rate (the rate at which a plate moves away from the axis) is 2 cm yr^{-1}, then the total rate of formation of new oceanic crust is twice this, i.e. 4 cm yr^{-1}, or 0.04 m yr^{-1}. At this rate, 1 m of spreading is accomplished, on average, in a period of $1 / 0.04 = 25$ years. So one new dyke can be expected about every

25 years. The East Pacific Rise is spreading four times faster, so if the dykes are the same width then they will be generated four times as often, i.e. about once in every six years.

Question 4.3 This trough is not really part of the median valley, which, as described in Section 2.3, is a feature produced by extensional faulting. In fact, this marks the Kane transform fault, which runs at a right angle to the trend of the spreading axis, allowing relative sideways motion between the sea-floor on either side. It was noted in Section 2.4 that transform faults and fracture zones are often expressed as clefts on the ocean floor. (The Kane transform fault offsets the median valley by about 130 km, and if you look carefully at Figure 4.9 you can see the beginning of the southward continuation of the median valley at about 45° W. A transform fault exists only between the two offset lengths of spreading axis, so to the east of this point the east–west trough is actually part of the Kane *fracture zone*.)

Question 4.4 To be consistent with the seismic data that show no indications of magma chambers and the model presented in Figure 4.5, the 'small flat magma bodies' in Figure 4.11 would be composed of a crystal mush rather than being entirely magma-filled. However, the possibility of bodies of crystal-free magma occurring locally and for limited periods of time cannot be ruled out.

Question 4.5 Figure 4.10(b) shows the median valley to be about 9 km wide. At a full-spreading rate of 25 mm yr^{-1}, the time interval between eruption in the centre of the valley and arrival at the valley walls must be:

$$\frac{9 \times 10^6 \text{ mm}}{25 \text{ mm yr}^{-1}} = \frac{9 \times 10^6 \text{ yr}}{25} = 360\,000 \text{ yr.}$$

[*Comment:* you could find the same value by dividing half the width of the valley by the half-spreading rate.]

Question 4.6 It is important to bear in mind that we have reported details from only one part of the Mid-Atlantic Ridge (20° N to 40° N), and one part of the East Pacific Rise (8° S to 10° N). However, if we assume that these are typical, then we can say that the Mid-Atlantic Ridge is characterized by eruptions (commonly of pillow lavas) at volcanic abyssal hills and along ridge-parallel fissures (to produce hummocky volcanic ridges). These eruptions occur on the floor of the median valley, especially towards its centre where they combine to produce an axial volcanic ridge within the valley, which subsides (or is faulted down) to the valley floor as spreading progresses. The abyssal hill volcanoes are up to a couple of kilometres in diameter, and the hummocky ridges are up to a few kilometres long. In contrast, eruptions on the East Pacific Rise seem not to build any discrete edifices (aside from major seamounts); instead, lava flows tend to be more sheet-like (rather than pillowed) and spread out for tens of kilometres.

Question 4.7 The minimum rate of creation of new crust in the Ontong–Java Plateau is calculated from its minimum likely volume (36 million km^3) and the maximum duration of time (3 million years) for its formation:

$$\frac{36 \text{ million km}^3}{3 \text{ million yr}} = 12 \text{ km}^3 \text{ yr}^{-1}.$$

The maximum rate is calculated from its maximum likely volume (76 million km^3) and the minimum duration of time (0.5 million years) for its formation:

$$\frac{76 \text{ million km}^3}{0.5 \text{ million yr}} = 152 \text{ km}^3 \text{ yr}^{-1}.$$

The rate we have calculated for the Ontong–Java Plateau (12–152 $km^3\ yr^{-1}$) could be comparable to the total global average rate of oceanic crust production of 16–26 $km^3\ yr^{-1}$ (if the true rate for the Ontong–Java Plateau lies close to its minimum possible rate of 12 $km^3\ yr^{-1}$); or it could be several times greater (if the true rate for the Ontong–Java Plateau lies in the upper part of its possible range).

Question 4.8 All it means is that in the long term, overlapping spreading centres stay put, but may migrate or oscillate over short distances about some mean position every few hundred thousand years.

(b) Rock samples dredged from the western limb of the overlapping spreading centre will have been derived from the magma lens beneath the topographic high immediately to the south; samples from the eastern limb would have come from the magma lens beneath the topographic high to the north. We should expect the samples to show some subtle differences in their chemical and mineralogical characteristics.

Question 4.9 Hydrostatic pressure tends to prevent the escape of dissolved gas from magma erupted under water, and any gas that does escape will be compressed into smaller vesicles than at atmospheric pressure. The predominance of lavas over volcanic ash in the lower parts of oceanic volcanoes can be explained by high hydrostatic pressures at the sea-floor which prevent gas escaping explosively. Chances of a 'steam explosion' are much reduced because the high pressures not only raise the boiling point of water but also inhibit the expansion of any steam that might form. [*Comment:* the sudden release of high pressure can sometimes have dramatic effects. For example, blocks of vesicular lava dredged from deep water off southern California spontaneously burst apart when exposed to atmospheric pressure on board ship – the area has become known as 'Popcorn Ridge' in consequence. However, vesicles are not commonly found in submarine lavas beneath about 500 m depth.]

CHAPTER 5

Question 5.1 (a) The *observed* measurements of heat flow across spreading axes differ increasingly from the *predicted* values as the ridge crest is approached. The theoretical curves assume that heat is transferred only by conduction. The observations show this to be less than predicted, so there must also be transfer of heat by convection of seawater within the crust.

(b) The shaded areas on Figure 5.6 represent the 'heat loss deficit', the *difference* between predicted conductive heat loss (broken line) and the observed conductive heat flow. This difference must be being transferred out of the oceanic crust by convecting seawater.

(c) Both curves have an exponential form, because increasing depth (Figure 2.13) and decreasing conductive heat flow (Figure 5.6) are closely related in models of cooling lithospheric plates.

(d) The fact that the four components of Figure 5.6 are so similar shows that this pattern must be typical of the world's oceans. On all four diagrams, observed heat flow matches prediction only from about 70 Ma onwards (cf. Figure 2.13), so in crust less than about 70 Ma a significant amount of heat transport can be inferred to be by hydrothermal circulation. Figure 3.6 suggests that oceanic crust up to about 70 Ma old occupies about one-third of the area of the world's oceans.

Question 5.2 (a) (i) The variation in the analyses of Table 5.1 enables us to be sure only that two constituents are added to basalt from seawater during hydrothermal metamorphism, namely MgO and H_2O. (ii) Similarly, we can be sure only of two constituents that are consistently leached from rocks, namely CaO and K_2O. There is some indication that SiO_2 is leached to varying degrees. The data in Table 5.1 suggest that Na_2O, TiO_2, iron oxides and MnO may be either added to the rocks or removed from them (cf. Section 5.2.2).

(b) The answer lies in fault scarps. Faulting in axial regions forms escarpments roughly parallel with ridge trends. Transform faults form scarps and clefts across the trend of ridges, with displacements of as much as several hundred metres, so that their lower parts will expose deeper layers of oceanic crust at the surface (Section 2.4 and Figure 2.16).

Question 5.3 (i) The heading for Table 5.2(b) states that hydrothermal solutions have a pH of around 4, and so must be more acidic than ordinary seawater which has a pH of around 8. (ii) They are also more reducing, because they contain sulphide (as H_2S) rather than sulphate (SO_4^{2-}) in solution, as well as more iron and manganese, which are more soluble in reducing than in oxidizing conditions.

Question 5.4 (a) If we assume a cracking front penetration rate of $3\ \mathrm{m\ yr^{-1}}$, it would take about $\frac{5000\ \mathrm{m}}{3\ \mathrm{m\ yr^{-1}}} = c.\ 1700$ years for the water to penetrate to a depth of 5 km, i.e. this is the time required for the rock at 5 km depth to have become cool and brittle enough to crack. Thus, the lifespan of both crystal mush zones and vent systems should be of the order of millennia (10^3 years), unless there is replenishment by fresh melt from below within that time.

(b) The mass of gabbro in a $10^{10}\ \mathrm{m^3}$ volume (density × volume) = $10^{10}\ \mathrm{m^3} \times 2500\ \mathrm{kg\ m^{-3}} = 2.5 \times 10^{13}\ \mathrm{kg}$. The heat liberated by crystallizing this volume (latent heat of solidification × mass) = $4.5 \times 10^5\ \mathrm{J\ kg^{-1}} \times 2.5 \times 10^{13}\ \mathrm{kg} = 11.25 \times 10^{18}\ \mathrm{J}$. At a rate of 2×10^8 W, which is $2 \times 10^8\ \mathrm{J\ s^{-1}}$, the lifetime would be

$$\frac{11.25 \times 10^{18}\ \mathrm{J}}{2 \times 10^8\ \mathrm{J\ s^{-1}}} = 5.6 \times 10^{10}\ \mathrm{s}.$$

Dividing this by the number of seconds in a year, we get the lifetime in years,

$$\frac{5.6 \times 10^{10}\ \mathrm{s}}{32 \times 10^6\ \mathrm{s\ yr^{-1}}} = 1.75 \times 10^3 \text{ years}$$

which (given the rough and ready nature of the assumptions we have made) we can approximate to 2000 years – so again we have a lifetime in the order of millennia, i.e. 10^3 years.

Question 5.5 The two calculated estimates in Question 5.4 were approximately 1700 and 2000 years, both an order of magnitude less than the estimated lifetime of the TAG mound. The calculation assumed no recharge of magma into the system, whereas if a fresh pulse (or continual trickle) of magma were supplied to the axial zone then the system could operate for longer. However, there is another important factor involved in prolonging the mound's life, discussed in the text, which does not require any extra heat input.

Question 5.6 You may have been tempted, by analogy with the conductive anomaly at the axis, to say that low heat flow represents convective upflow. However, off axis the situation is different, because here convection redistributes heat *within* the crust, and because the circulation is mostly sealed in, it does not actually transport much heat out of the crust. *High* conductive heat flow is measured in *upflow* regions, whereas the conductive heat flow is *lower* in downflow regions. This is because downflowing water carries heat away, cooling the rock and reducing its heat flow. Upflowing water brings with it the heat which it gained elsewhere and warms the rock up, increasing its heat flow. Note that there is still an overall deficit compared with the theoretical conductive heat flow, because some heat is transported out of the crust (into the seawater) by convection.

Question 5.7 The seismic P-wave velocities of unmetamorphosed and metamorphosed oceanic crustal rocks are not sufficiently different to enable definite discrimination to be made between them using seismic information alone. For example, gabbros are indistinguishable from amphibolite grade rocks and young basalts are indistinguishable from greenschist grade rocks. [*Comment:* Figure 5.17 shows only data for seismic compressional wave (P-wave) velocities, but similar overlaps occur between shear wave (S-wave) velocities.]

Question 5.8 (a) Substituting values into Equation 5.1:

$$F = \frac{2 \times 10^{20}\,\text{J yr}^{-1}}{4.2 \times 10^{3}\,\text{J kg}^{-1}\,°\text{C}^{-1}\,(300\,°\text{C} - 2\,°\text{C})} \approx 1.6 \times 10^{14}\,\text{kg yr}^{-1}$$

(b) The total mass of ocean water divided by the rate calculated in (a) gives:

$$\frac{1.4 \times 10^{21}\,\text{kg}}{1.6 \times 10^{14}\,\text{kg yr}^{-1}} \approx 9 \times 10^{6}\ \text{years.}$$

Question 5.9 Table 5.3 shows that the degree of hydration (the amount of H_2O incorporated in the rock) in sea-floor weathering is less extreme than in hydrothermal alteration; also, that sea-floor weathering causes potassium to be added to the rocks; and that oxidation causes Fe_2O_3 to increase considerably relative to FeO. The other effects (alteration of glass and feldspars, deposition of manganese oxide) are less obvious in Table 5.3.

Question 5.10 Figure 5.7 shows that the isotherms are very close together for temperatures less than *c*. 300 °C. The thickness of crust in which the temperature range is within the stability field of chlorite is much greater than that appropriate to zeolite stability. That is why chlorite is so much more abundant.

Question 5.11 (a) 1300 p.p.m. magnesium in 10^{14} kg yr^{-1} of water is $1300 \times 10^{-6} \times 10^{14}$ kg yr^{-1} of magnesium = 1300×10^{8} kg yr^{-1} = 1.3×10^{11} kg yr^{-1}.

(b) The answer to our simple calculation is the same as for the river flux. Hydrothermal systems must provide the *major* sink for magnesium in the oceans: the concentration of magnesium in seawater is kept constant because a quantity equal to all the magnesium supplied annually by rivers is removed in hydrothermal systems each year.

Question 5.12 (a) False. Sea-floor weathering will occur wherever seawater at bottom temperatures is in contact with oceanic crustal rocks.

(b) This is nonsense: sea-surface temperatures cannot affect hydrothermal processes.

(c) True. The first sediments to accumulate on igneous crust are deposited at ridge crests and are enriched in Fe and Mn (and some other metals) discharged at hydrothermal vents (Figures 5.4 and 5.5). Any sediments deposited thereafter will overlie these metal-rich sediments.

(d) False. Looking at Figure 5.7 it is likely that only the uppermost layer of rubbly lava, layer 2A, is saturated with cold seawater. Below that, water is beginning to heat up in downflow zones.

(e) True. The only source of ^{3}He of any consequence is mantle outgassing from ocean-ridge axes. Although the ^{3}He eventually escapes to the atmosphere, its highest concentrations will be in the oceans, though it is diluted all the time by ^{4}He, which is being produced wherever rocks contain uranium or thorium.

CHAPTER 6

Question 6.1 (a) The most important single factor controlling the volume of water in the oceans is the balance between the freezing and melting of ice, especially in the polar ice-caps, but also in mountain glaciers.

(b) A 1 °C increase in average temperature would cause the volume of the oceans to increase by a factor of 2.1×10^{-4} or 0.021%. Using the average depth of the oceans of 3.7 km, and neglecting any resulting change in surface area (which will be very small in comparison with the total surface area), an average temperature rise of 1 °C throughout the oceans would cause the depth to increase by a factor of 2.1×10^{-4}, due to thermal expansion of the water. This is a rise of:

$$3.7 \times 10^{3}\,\text{m} \times 2.1 \times 10^{-4} = 7.8 \times 10^{-1}\,\text{m} \approx 0.8\,\text{m}$$

(to one significant figure, which is all that we can justify).

So, ignoring ice-sheets, global temperature changes can cause sea-level changes of the order of a few metres. This does not seem much in relation to the depth of the oceans, but when we are considering changes of the order of millimetres (see the text that follows), then temperature changes could become an important factor.

(c) The depth to which oceanic crust will sink on cooling after it spreads away from a constructive plate margin depends upon its age (Figure 2.13) and not upon the distance it has moved since its formation. It follows that oceanic crust produced at a fast-spreading axis will have travelled further from the ridge before it sinks to a given depth. A fast-spreading ridge will therefore generally be broader and higher than a slow-spreading ridge. So, other things being equal, the ocean basin will on average be shallower and more water will be displaced onto continental shelves worldwide (cf. answer to Question 2.11).

Question 6.2 Figure 6.5(a) shows that the overall (secular) rise in sea-level for the 70-year period 1890 to 1960 was from about 20 mm to 150 mm on the vertical scale, a total rise of 130 mm over 70 years, giving an average rate of rise of 1.8–1.9 mm per year.

Question 6.3 (a) Ten-day averages were used because this is the time taken to complete the satellite's near-global coverage. Averaging over this period enables tidal effects and short-term longitudinal variations in sea-level to be removed.

(b) Relative to the (arbitrary) datum on the vertical scale, the best-fit sea-level line is at about −7 mm at the start of 1993 and at about + 5.5 mm at the start of 1995. This is a rise of about 12.5 mm over two years, which works out as an annual rate of rise of about 6.25 mm yr^{-1}. Because of the noise-like variation in the original curve, it is realistic to quote the rate simply as approximately 6 mm yr^{-1}.

Question 6.4 Sea-level is still rising within a belt extending from southern Britain through the North Sea and into the Netherlands and northern Germany. However, it is falling around most of Scandinavia, because this region is still rebounding isostatically from the release of the ice-loading after the most recent glaciation. This rebounding has outpaced the general rise in sea-level caused by ice-melting.

Question 6.5 (a) Zero. This is because $^{18}O : ^{16}O$ of sample and standard would be identical, and therefore the top line of the fraction defining $\delta^{18}O$ would be zero.

(b) Equation 6.1 tells you that a positive $\delta^{18}O$ value means the $^{18}O : ^{16}O$ ratio in the sample must be greater than that in the standard. It therefore indicates enrichment in ^{18}O in the sample relative to the standard (VSMOW in this case).

(c) A negative $\delta^{18}O$ value must by the same argument indicate relative depletion of ^{18}O.

(d) As organisms growing in cold water tend to incorporate more of the heavy ^{18}O isotope into their skeletons, the $^{18}O : ^{16}O$ ratio must be higher in organisms growing in cold water. We should therefore expect to find higher $\delta^{18}O$ values in the skeletons of such organisms.

(e) Evaporation enriches the ^{16}O isotope in water vapour, relative to the water left behind. There is therefore less ^{18}O in ice than in seawater and $\delta^{18}O$ values should be negative in polar ice.

Question 6.6 (a) It seems logical that the temperature of surface water varies much more than that of water in the deep oceans, where the variation in temperature throughout the Quaternary is unlikely to have exceeded a few degrees, as at the present day. Because the oxygen-isotope ratio in calcareous skeletal remains depends partly on the proportion of ^{16}O that is incorporated into polar ice-caps and partly on the temperature of the water in which the organisms grow, this second variable is largely eliminated by measuring isotope ratios of remains of organisms that grew where there was very little temperature variation. Benthic foraminiferans fulfil this criterion. [*Comment:* In fact, surface water temperatures at low and mid-latitudes were probably not much different during glaciations, except near the fringes of the ice, and planktonic foraminiferans (which are more abundant than benthic varieties) have also been used with apparent success in tropical regions. However, it seems that, in the Cretaceous (150–65 Ma), bottom waters were warmer and the oceans lacked the strong temperature contrast between surface waters and bottom waters that characterize today's oceans.]

(b) From Figure 6.11, and with the information that a 0.1 per mil change in the $\delta^{18}O$ value corresponds to a 10 m change in sea-level, it is possible to deduce (i) a rise of about 100 m for the last post-glacial interval, and (ii) a slightly higher rise (about 120 m) for the preceding interval.

(c) (i) The last major fluctuation in sea-level was not especially noteworthy because the last glaciation was not significantly different from its predecessors. (ii) The rises in sea-level that follow glaciations are evidently more abrupt than the falls, which seem to be characterized by oscillations superimposed on an overall downward trend.

Question 6.7 (a) The first definite signs of ice-transported debris in the sediments appeared in the late Oligocene, and the surface water temperatures were then of the order of 7 °C.

(b) There is a fairly abrupt steepening of the downward trend in the mid-Miocene, and this could reasonably be correlated with an acceleration of the growth of the Antarctic ice-sheet.

Question 6.8 If the net rate of water loss of $3.25 \times 10^3\,\text{km}^3\,\text{yr}^{-1}$ were maintained, since the initial volume of water in the Mediterranean is $3.75 \times 10^6\,\text{km}^3$, the time taken for it to dry out would be:

$$\frac{3.75 \times 10^6\,\text{km}^3}{3.25 \times 10^3\,\text{km}^3\,\text{yr}^{-1}} = 1.15 \times 10^3 \text{ years} \approx 1000 \text{ years}$$

Question 6.9 The thickness of evaporites with the specified volume and surface area would be:

$$\frac{6.5 \times 10^4\,\text{km}^3}{2 \times 10^6\,\text{km}^2} = 3.25 \times 10^{-2}\,\text{km, or } 32.5\,\text{m}$$

(which, given the uncertainties, and the fact that the density was quoted to only one significant figure, we should quote as ~30 m).

Question 6.10 (a) This part of the question is designed to ensure that you recognize the general correlation: the higher the δ^{18}O value, the more ice there was in glaciers and ice-caps, and hence the lower the sea-level. Average global temperatures were probably lower also (cf. Section 6.2.4) which would tend to reduce the rate of evaporation, but nevertheless evaporite deposition occurred during the periods of lower sea-level, when the Mediterranean was isolated and so could dry up due to evaporation.

(b) Each of the numerous peaks and troughs in the raw data indicates fluctuating ice volumes and hence fluctuating sea-levels. Short-term rises and falls of global sea-level would provide the necessary inundations to replenish the supply of seawater in the Mediterranean and produce a new 'crop' of evaporites.

(c) The time-scale varies somewhat: there are about 10 peaks in the interval 5.89–5.35 Ma, and about another 10 in the interval 4.77–3.88 Ma. The total time interval in the first case is about half a million years, in the second about 900 000 years. The time-scale of fluctuations thus varies from about 50 000 to 100 000 years. This approximates the time-scale of fluctuations in Figure 6.11. However, it is crucially important when making such comparisons to bear in mind the resolution of the sampling. In Figure 6.15, the sampling interval was every 70 cm of sediment, corresponding to about 20 000–25 000 years; in Figure 6.11, sampling was about every 10 cm, corresponding to about 5000–6000 years. Thus, the record in Figure 6.11 resolves details that are lost in Figure 6.15.

(d) The Messinian salinity crisis lasted from about 5.5 Ma ago (when the overall lowering of sea-level and tectonic uplift combined to isolate the Mediterranean) to about 4.8 Ma ago (when the barrier at Gibraltar was finally submerged), a total of 0.65 Ma. However, there was an interval between about 5.3 and 5.2 Ma ago when normal marine conditions intervened between the two periods of evaporite deposition.

(e) Falls in sea-level were no longer able to isolate the Mediterranean once the barrier at Gibraltar was opened finally, because the connection had become too deep, probably mainly through tectonic effects, possibly also by erosion from inflowing water.

(f) In Section 6.2.5 you read that growth of the Antarctic ice-sheet, the world's major repository of ice, began in earnest in the mid-Miocene (*c.* 10 Ma ago), whereas significant growth of Northern Hemisphere ice-caps did not begin until *c.* 3 Ma ago. It is likely that most of the sea-level fluctuations in the Miocene and later times were caused by changing ice volumes in Antarctica.

Question 6.11 Remarkably enough, no. The distribution of surface isotherms (lines of equal temperature) is clearly not the same in detail in both diagrams, but the zone of warm water in equatorial and tropical latitudes (up to about 30 °N in Figure 6.16) was apparently not much narrower at the height of the glaciation than it is now. Further north, however, the isotherms were formerly much more crowded together as a result of the southward expansion of the high-latitude cold belt. It is the mid-latitude climate belt that was compressed, not the tropical belt. [*Comment:* for this reason, tropical planktonic foraminiferans can be used, as well as benthic species, in oxygen-isotope estimates of sea-level; see answer to Question 6.6(a).]

Question 6.12 There is no obvious evidence of any such control – indeed, rather the reverse: there were regressions after the Ordovician and Permo-Carboniferous glaciations, when polar ice-caps shrank. The very extensive transgression in the Cretaceous cannot be associated with the decline of any glaciation – and the ensuing regression (*c.* 70 Ma ago) could not have been caused by the growth of polar ice-caps, either. These major transgressions must have resulted from plate-tectonic processes, or perhaps the development of large igneous provinces.

Question 6.13 (a) True. The rise in sea-level since the last glacial maximum is at least 90 m (Figure 6.9), and an extra 60 m could be expected if the rest of the ice melted (Section 6.2.4).

(b) False. The *TOPEX/Poseidon* data presented in Section 6.2.2 show an annual rise in sea-level of approximately 4 mm yr^{-1} over a two-year period (Figure 6.7). This coincides with a period of increase in global average sea-surface temperature (Figure 6.8), but this could be only a temporary excursion related to an El Niño–Southern Oscillation (ENSO) event rather than a long-term trend. A longer time-base of altimetrically derived global sea-level determinations is required to demonstrate any long-term trend, and even then the exact causal link between global warming and rising sea-level would remain to be established.

(c) False. *Guembelitria* has helped to establish when open water conditions developed all round Antarctica with a probable wind-driven surface current, but it is a planktonic organism and cannot be satisfactorily used to determine when the current became fully developed *at depth*.

(d) False. We emphasized at the end of Section 6.2.6 that such sequences *can* be formed at these stages, but this does not invariably happen.

CHAPTER 7

Question 7.1 The answer must be no, especially after reading Chapter 5. Considerable quantities of elements have been added to the crust *en route*, e.g. magnesium, Mg, and potassium, K. Large amounts of other elements have been removed, notably calcium, Ca. These and other elements have also been redistributed, so that, for example, it is likely that large amounts of iron and manganese (and perhaps copper and other ore metals) have been removed from within the crust and redeposited near or on the surface, sometimes in concentrations of economic importance. Above all, huge volumes of water have become fixed in the crust, in hydrous minerals such as clays, zeolites and chlorite. These changes occur most rapidly nearest the spreading axes.

Question 7.2 You know from Section 5.2 that during hydrothermal reactions at spreading axes, Mg^{2+} is removed from seawater and combined into new minerals within the hydrothermally altered rock. With a shorter length of spreading axis in the Permian, we could expect a globally reduced rate of hydrothermal transfer of Mg^{2+} from seawater to rock. Unless the rate at which Mg^{2+} was being supplied to the oceans (by rivers) was somehow reduced by an appropriate proportion, we would expect the Mg^{2+} content of seawater to increase, which would therefore favour the precipitation of Mg salts in settings where evaporites were forming. Similar arguments apply to the SO_4^{2-}, the proportion of which reduced to sulphide (Table 5.2) would be smaller. There would be more SO_4^{2-} in seawater and more of the Mg^{2+} would be precipitated as sulphate. When Pangea broke up, the total length of spreading axis would have increased to something like its present-day amount, increasing the rate of removal of Mg^{2+} and SO_4^{2-} once more.

Question 7.3 (a) 1.4×10^{21} kg water with 35 g salts per kg means that the total quantity of salts is:

$$1.4 \times 10^{21} \times 35 \times 10^{-3}\,\text{kg, i.e.} \sim 5 \times 10^{19}\,\text{kg}$$

(b) The Mediterranean evaporites represent:

$$\frac{4 \times 10^{18}}{5 \times 10^{19}} \times 100$$

or 8% of the salt in the world oceans.

Question 7.4 (a) $50 \times 10^6\,\text{km}^3$ is $50 \times 10^{15}\,\text{m}^3$.

This is $50 \times 10^{15} \times 1025 = 5.125 \times 10^{19}\,\text{kg} \approx 5.1 \times 10^{19}\,\text{kg}$.

The total mass of water in the ocean is 1.4×10^{21} kg, so the glacial fluctuation is:

$$\frac{5.1 \times 10^{19}}{1.4 \times 10^{21}} \times 100 \approx 3.6\%$$

(b) As there is less *water* in the oceans during glacial periods, we might expect the average salinity to be greater. [*Comment:* the freezing of seawater produces freshwater ice and expels cold supersaline water. This is important, because if frozen seawater ice were salty, then ice formation would not affect seawater salinity.]

Question 7.5 (a) Less calcium entering a steady-state ocean must eventually result in less calcium being removed from it (or more being added to it by another means). There are various ways in which this is likely to be achieved: decreased precipitation of calcium carbonate to form skeletons and shells of marine organisms; increased dissolution of biogenically precipitated $CaCO_3$; and increased supply of calcium from hydrothermal activity.

(b) More magnesium entering a steady-state ocean must eventually result in more magnesium being removed from it. Increased magnesium concentration in sediment pore waters would result in increased removal by diagenetic reactions (Figure 7.2); more magnesium would also be removed from seawater during hydrothermal circulation (cf. Question 7.2).

Question 7.6 The answer is given by the expression:

$$\frac{1\,\text{m}}{6 \times 10^{-7}\,\text{m}\,\text{s}^{-1}} \text{ or } 1.67 \times 10^{6}\,\text{s}.$$

That works out at just over 19 days, corresponding to a rate of about 5 cm per day, which is consistent with rates to be expected in the downflow zone of hydrothermal systems, as suggested in Section 7.2.

Question 7.7 The reason is that the concentration of calcium in hydrothermal solutions is much greater than that in river water (although you have not been provided with data, you know that freshwater is not saline). A small amount of concentrated solution makes a proportionately greater contribution than a large amount of a dilute solution.

ACKNOWLEDGEMENTS

The Course Team wishes to thank the following: Dr. Martin Angel and Dr. Andy Fleet, the external assessors; Mr. Mike Hosken for advice and comment on the whole Volume; Captain Iain F. Kerr and Dr. Charles Turner for their help in the compilation of Chapters 1 and 6 respectively; the late Dr. Graham Jenkins and Dr. Sandra Smith for information on specific points.

For this second edition, we wish to thank also the external assessors Dr. Martin Angel (again) and Dr. Bramley Murton; Dr. Glen Middleton for his help with Section 1.1.1; various students and tutors (too numerous to name) for useful comments and suggestions on the first edition; Dick Carlton for technical help; and colleagues in other institutions who allowed their brains to be picked.

The structure and content of the Series as a whole owes much to the experience of producing and presenting the first Open University course in Oceanography (S334) from 1976 to 1987. We are grateful to those people who prepared and maintained that Course, to the tutors and students who provided valuable feedback and advice and to Unesco for supporting its use overseas.

Grateful acknowledgement is also made to the following for material used in this Volume:

Figure 1.1 Ancient Art and Architecture Collection; *Figure 1.2* N. J. W. Thrower; *Figure 1.3* Naval Museum, Madrid, and Arixiu Mas, Barcelona; *Figure 1.4(a)* E. Linklater (1972) *The Voyage of the Challenger*, John Murray Ltd. by permission of Rainbird Publishing Group; *Figure 1.4(b)* Institute of Oceanographic Sciences/NERC; *Figure 1.8* K. K. Turekian (1976) *Oceans,* Prentice-Hall Inc.; *Figure 1.11* NASA; *Figure 1.13(b)* C. Flewellen *et al.* (1993) in *Electronics and Communication Engineering Journal*, April; *Figure 1.14* N. G. Dukov, Institute of Oceanography, Varna, Bulgaria; *Figures 1.15, 1.18, 2.26–2.28* US Dept. of Commerce, National Oceanic & Atmospheric Administration (NOAA), National Geophysical Data Center, and World Data Center for Marine Geology and Geophysics; *Figure 1.19* courtesy of the National Remote Sensing Centre, Farnborough; *Figure 1.20* National Science Foundation; *Figures 1.22, 4.8(b) and 5.9(c)* Woods Hole Oceanographic Institution; *Figure 2.1* X. Le Pichon *et al.* (1973) *Developments in Geotechnics 6: Plate Tectonics,* Elsevier; *Figures 2.8 and 2.10* C. A. Burk and C. L. Drake (1974) *The Geology of Continental Margins*, Springer-Verlag; *Figure 2.12* H. D. Needham and J. Francheteau (1974) in *Earth and Planetary Science Letters,* **22,** No. 1, North Holland Publishing; *Figure 2.13* J. G. Sclater and D. P. McKenzie (1973) *Geological Society of America Bulletin,* **84,** 10, Geological Society of America; *Figure 2.18* Deborah K. Smith, Woods Hole Oceanographic Institution; *Figure 2.19* A. S. Laughton (1959) in *New Scientist,* **1**, New Science Publications; *Figure 2.20* B. G. Heezen and C. D. Hollister (1971) *The Face of the Deep,* Oxford University Press; *Figure 2.22* G. B. Dalrymple *et al.* (1973) in *American Scientist,* **61**; *Figure 2.24* T. H. Dixon and M. E. Parke (1983) in *Nature,* **304**, Macmillan; *Figure 3.3(a)* E. Bonatti (1985) in *Nature,* **316**, Macmillan; *Figure 3.4* E. T. Degas and D. A. Ross (eds) (1969) *Hot Brines and Recent Heavy Metal Deposits in the Red Sea,* Springer-Verlag; *Figure 3.6* J. G. Sclater and B. Parsons (1981) in *Journal of Geophysical Research,* **91,** B12; *Figure 3.7* R. G. Gordon and J. M. Gurdy (1986) in *Journal of Geophysical*

Research, **91**, B12; *Figure 4.4(a)* R. S. Detrick *et al.* (1987) *Nature,* **326**, Macmillan; *Figures 4.4(b) and 4.21* D. R. Toomey *et al.* (1990) in *Nature*, **347**, Macmillan; *Figures 4.6 and 4.8(a)* D. A. Rothery; *Figure 4.7(a)–(c)* J. G. Moore (1975) in *American Scientist,* **64**; *Figure 4.7(d)* J. R. Heirtzler and W. B. Bryan (1975) in *Scientific American*, **233**, copyright © 1975 W. H. Freeman and Co.; *Figure 4.9* D. K. Smith and J. R. Cann (1990) in *Nature*, **348**, Macmillan; *Figures 4.10 and 4.11* D. K. Smith and J. R. Cann (1993) in *Nature*, **365**, Macmillan; *Figure 4.13* K. C. MacDonald *et al.* (1988) in *Nature*, **335**, Macmillan; *Figures 4.14 and 4.15* R. H. Haymon *et al.* (1993) in *Earth and Planetary Science Letters*, **119**; *Figure 4.16 K. C. MacDonald et al.* (1989) in *Geology*, **17**; *Figures 4.18, 4.19(a) and 4.20* K. C. MacDonald and P. J. Fox (1983) in *Nature,* **302**, Macmillan; *Figure 4.19(b)* K. C. MacDonald *et al.* (1984) in *Journal of Geophysical Research,* **89**, B7; *Figure 4.22* J.-C. Sempéré *et al.* (1990) in *Nature*, **344**, Macmillan; *Figure 4.23(a)* J. A. Whitehead (Jnr.) *et al.* (1984) in *Nature,* **312**, Macmillan; *Figure 4.24* G. P. L. Walker (1975) in *Nature,* **255**, Macmillan; *Figure 4.26* S. Thorarinsson; *Figure 4.27(b)* D. J. Fornari *et al.* (1984) in *Journal of Geophysical Research,* **89**, B13; *Figure 5.1* John Edmond/Woods Hole; *Figure 5.2* Robert Hessler/Woods Hole; *Figure 5.3* J. B. Wright; *Figure 5.6* R. N. Anderson (1981) in *Nature,* **292**, Macmillan; *Figure 5.15* E. T. Baker *et al.* (1989) in *Journal of Geophysical Research*, **94**, B17; *Figure 5.16* courtesy of B. Chadwick and B. Embley; *Figures 5.17 and 6.1* R. N. Anderson (1986) *Marine Geology,* John Wiley and Sons; *Figure 6.2* B. J. Skinner and K. K. Turekian (1973) *Man and the Oceans,* Prentice-Hall, first printed and by permission of J. Ben Carsey (1950) in *Bulletin of the American Association of Petroleum Geologists*, **34**, AAPG; *Figure 6.3* D. G. Jenkins; *Figure 6.4* D. G. Jenkins and J. P. Kennett; *Figure 6.6* courtesy of Lee-Leung Fu, C. J. Koblinsky, J. F. Minster and J. Picaut, and Jet Propulsion Laboratory; *Figures 6.7 and 6.8* R. S. Nerem (1995) in *Journal of Geophysical Research*, **100**, C12; *Figure 6.9* F. P. Shepard (1963) *Submarine Geology,* Harper & Row, copyright © 1948, 1963 Francis P. Shepard; *Figure 6.11* N. J. Shackleton and N. D. Updyke (1973) in *Quaternary Research,* **3**, Academic Press; *Figure 6.12* J. P. Kennett *et al.* (eds) (1974) *Initial Reports of the DSDP,* Vol. 29, US Government Printing Office; *Figure 6.13* S. G. Smith; *Figure 6.14* Guy Billout; *Figure 6.15* D. A. Hodell *et al.* (1986) in *Nature,* **320**, Macmillan; *Figure 6.16* W. L. Gates and L. Mintz (eds) (1975) *Understanding Climatic Change,* App. A, National Academy of Sciences; *Figure 6.17* adapted from A. Hallam (1984) in *Annual Review of Earth and Planetary Science,* **12**, copyright © 1984 Annual Reviews Inc.

INDEX

Note: page numbers in italic refer to illustration captions